Trends in Geography
An Introductory Survey

Trends in Geography
An Introductory Survey

Edited by

RONALD U. COOKE

LECTURER IN GEOGRAPHY, UNIVERSITY COLLEGE LONDON

and

JAMES H. JOHNSON

READER IN GEOGRAPHY, UNIVERSITY COLLEGE LONDON

PERGAMON PRESS

Oxford · London · Edinburgh · New York
Toronto · Sydney · Paris · Braunschweig

THE QUEEN'S AWARD
TO INDUSTRY 1966

Pergamon Press Ltd., Headington Hill Hall, Oxford
4 & 5 Fitzroy Square, London W.1

Pergamon Press (Scotland) Ltd., 2 & 3 Teviot Place, Edinburgh 1

Pergamon Press Inc., Maxwell House, Fairview Park, Elmsford, New York 10523

Pergamon of Canada Ltd., 207 Queen's Quay West, Toronto 1

Pergamon Press (Aust.) Pty. Ltd., 19a Boundary Street,
Rushcutters Bay, N.S.W. 2011, Australia

Pergamon Press S.A.R.L., 24 rue des Écoles, Paris 5ᵉ

Vieweg & Sohn GmbH, Burgplatz 1, Braunschweig

Printed in Great Britain by A. Wheaton & Co., Exeter

G
62
T 7
1968

08 006674 7 (flexicover)
08 006675 5 (hard cover)

CONTENTS

v

PART III. APPLIED GEOGRAPHY

PREFACE

THIS collection of essays had its starting point in a conference for geography teachers organized by the University of London Institute of Education and held at University College London in the summer of 1968. It attempts to examine some of the recent trends in geography and to suggest ways in which these tendencies may be important for teachers and students. It has been our aim to produce a short and inexpensive book. As a result, all our contributors have had to accept the necessity of compressing their work into a relatively small number of words, with the high degree of selection and generalization which this process involves.

For the same reason it has not been possible to refer to every field of study found in university departments of geography. The omission of a particular topic does not mean that work in it is without value; but we have tried to select those branches of the subject in which the interest of both research workers and students is lively and currently growing. Some of the interests described here represent a radical break with tradition, and at present do not form part of many geographical studies, at least as they are taught in schools. Others are established parts of geographical syllabuses, with trends which are often logical developments of familiar themes.

Each contribution gives a personal view of a particular field of study. We have thought it best not to impose a single and possibly restrictive methodological framework on our contributors, since one man's frontier of research may be another's base-camp. Yet it will be clear to readers that some recurring themes can be discerned through the counterpoint provided by the pre-occupations of individual geographers. The quest for precision, the adoption of approaches emphasizing the complex systems that surround individual geographical phenomena, the reliance on quantitative, and especially statistical, techniques—all these important features emerge in chapter after chapter.

Techniques of study are changing more rapidly in modern geography than at any previous time in the subject's history. As a result there is a great need for a dialogue between research workers and those being admitted to the mysteries of the subject. Teachers provide the necessary link; and it is dangerous for the vitality and future health of geography that some teachers find current developments either incomprehensible or unacceptable. As the engines of research move at increasing speeds along new lines, some of the carriages carrying teachers and students are in danger of becoming detached.

This book attempts to help rectify such a dangerous situation by discussing current work in as simple a manner as possible. It does not try to restate the benefits of teaching geography in schools, nor does it seek to advance another new frame of reference for research in geography. Rather, this book adopts the less ambitious strategy of occupying the middle ground: it is more concerned with the logistics of communication than with the advance of the battle-front.

As editors we have attempted to eliminate jargon from this book. As a result we hope that its contents will at least be comprehensible to teachers, first-year undergraduates and intelligent sixth-formers in Britain, and their equivalents elsewhere. The References and Further Reading sections are designed to serve several purposes. They list important research contributions to the literature which have set trends; they also include general and more readily accessible references concerned with the matters under discussion. In addition, we have tried to provide material in these sections which will enable those who are interested to read further. The references are classified in each section according to the subdivisions in the text; where a reference is used several times in the same chapter it is listed only once.

We owe a great debt of gratitude to our contributors, who have not only willingly accepted the occasional need to render their sonorous and subtle phrases into much balder statements, but have produced drafts of their chapters to a very tight schedule. We are also most indebted to Mr. K. Wass, senior cartographer in the Department of Geography at University College London, for his skilful assistance in producing the illustrations, and to our typists, Mrs. E. Jamieson, Mrs. F. Barton, and Miss D. Parker, for the speed and efficiency with which they have dealt with the manuscript.

University College London   RONALD U. COOKE
                                                         JAMES H. JOHNSON

ACKNOWLEDGEMENTS

THE authors wish to express their thanks to the following for permission to reproduce illustrations:

Professor E. F. Brater, University of Michigan, for Fig. 7; Professor J. Büdel, University of Würzburg, for Fig. 3; Professor P. Haggett, University of Bristol, for Fig. 11; Professor C. D. Ollier, University of Papua and New Guinea, for Fig. 2; Professor J. D. Ovington, The Australian National University, for Fig. 5; Professor D. M. Smith, Southern Illinois University, for Fig. 19.

INTRODUCTORY

CHAPTER 1

THE COURSE OF GEOGRAPHICAL KNOWLEDGE

W. R. MEAD

THE contemporary study of geography contains two paradoxes and in them are rooted both the problems and opportunities of the subject. The first paradox is that at a time which may be fairly claimed as the climax of general interest in geography, adjustments in attitude are demanded which must necessarily test the faith and resilience of its followers. The second paradox, which is partly related to the first, is that an increasing unity of approach to the subject is accompanied by an unparalleled diversification of activity within it. The resulting shifts of emphasis produce stresses and strains. The degree of their impact necessarily varies from area to area within the field of study. For many, the new experiences are exhilarating; for some, they are distressing. If the new problems that they pose are not of universal appeal, the possible new solutions to old problems that they offer arouse considerable attention. In the changes that are taking place new areas of research will be opened (some on the score of methodology alone), some existing areas of research are likely to receive a new stimulus, and other traditional lines of investigation will be correspondingly subordinated.

As an academic subject, geography has never claimed more adherents or more attention than at the present time. The number of students engaged in it (as measured by school and university examinations alone), the membership of its professional societies, and the number and variety of appointments in which geographers find outlets for their particular aptitudes, have never been greater. The adjustments which this body of disciples is called upon to make spring from both general and particular causes. They are general to the extent that the course of knowledge as a whole determines the course of its parts; all disciplines are affected by the profound changes that have taken place in the means of multiplying knowledge and the methods of diffusing it. They are particular in that there are changes in the content and methodology of geography that are specific, or nearly specific, to the subject.

Geography has always had close relations with a range of cognate subjects. In the past, links were principally with surveying, history, geology, botany, and anthropology. When economics emerged as a university subject, a close relationship was also struck with it. Geography in the University of London, for example, entered the modern examination syllabus by the way of the B.Sc. (Economics) degree. As branches of the subject began to distinguish themselves, adjectives proliferated to describe them. The established liaisons remain, with geography making a growing contribution to kindred disciplines as well as gathering ideas and material from them. Simultaneously, new associations are developing, though the resulting branches of the subject are less frequently identified by adjectives than formerly. The statistician, sociologist, logician, psychologist, hydrologist, engineer, architect and, of course, planner, are all courted by the geographer—and sometimes court him. Again it must be emphasized that the search for and fulfilment of new relationships are not unique to the geographer. They are characteristic of the age.

At the same time as individual geographers or schools of geographers have been establishing these new liaisons, a common and increasingly well-identified methodology has been evolving. In part, this derives from the social scientists; in part it is the contribution of geographers themselves. The methodology in particular identifies the present decade as a time of change in the accepted structure of the geographer's field of learning. Geography as the study of the unique—the exceptionalist view—is rapidly yielding to other approaches. It is challenged by the functional appreciation of geographical data and by the constructional approach to them. It is confronted by the emphasis on relative rather than on absolute considerations.

The steady restructuring of much of the subject promises to yield new systems of knowledge. Although the new systems look to the sign language of the mathematician as unambiguous and facilitating manipulation, they have also prompted the development of a new vocabulary. An immense variety of new words and phrases have been and are being conceived to meet the needs of the situation. As with most subjects, geography has accumulated its language from many sources, but hitherto its vocabulary has been of an essentially concrete character. The contemporary vocabulary includes an increasing number of abstract words and phrases. They derive principally from the social sciences (from sociology and economics in particular); partly from mathematics and statistics, and to a lesser extent from the physical sciences. Despite its scientific source, the torrent of new words and concepts that has invaded geographical literature is curiously poetic in sound. Paradigm and parameter, matrix and linkage, ecosystem and trend surface are of the stuff that dreams, as well as revolutions, are made. In the midst of the torrent, a few rock-like words resound—precision, perception, and prediction are among them. Even these are abstract, but they have a comfortingly familiar ring.

Above all, contemporary geography aims at precision—precision through measurement. For some decades, geographers have been groping towards the fuller and more effective use of statistical materials and methods. The slowly changing approach to work in the field, the increasing refinement of map-making from statistics, the steady realization that the graphical representation of data had its own geographical meaning—all anticipated the impulse towards quantitative expression. The ready reckoner and slide rule yielded to the punch-card and the data bank, the desk computer and the electronic brain. Geography has leapt forward to employ to the full the mechanical equipment of the mid-twentieth century, for the bulk of its raw materials lend themselves to ready use of quantitative analytical techniques. A new scale of operations becomes evident in the process. It is illustrated by the Canadian Land Inventory, by Swedish experiments to establish a kilometre grid record for the entire country, and by the pilot scheme of a similar character initiated for East Anglia by the cartographic unit of the Royal College of Art. A similar transformation has taken place in the pictorial recording of geographical features. It is only a generation since extensive large-scale airphotographic coverage began to be available and since photogrammetric devices transformed survey. The introduction of satellites has given new perspectives to airphotography and, in turn, these have an immediate effect upon the production and interpretation of maps (Bird and Morrison, 1964).

The machines that the geographer has at his disposal are part servant, part master. In order to use them he must either learn new skills or employ the aid of technical experts. Most demanding in the battery of equipment is the computer, which requires appropriately programmed material. Some geographers become programmers in their own right and computer mapping is already a distinct field of study. In addition to transforming the speed with which information can be made available, computers may also change the form of its availability. While the computer is an enormous aid to the development of the subject, computerization of material for its own sake emerges as an occupational hazard.

A further consequence of precision through measurement is that geography inclines increasingly towards quantitative expression. This development raises one of the academic geographer's most vexacious problems. It may be fairly argued that an elementary knowledge of statistics and mathematics should be a primary component in the training of the geographer. Whether this training should take place in school or university or both remains undecided. In the short period it is evident that to demand more than an elementary knowledge of statistics and mathematics as a minimum requirement for entry into university might have a serious effect upon recruitment. While it is clear that Advanced Level mathematics or statistics are a highly desirable qualification for a potential geographer, it is equally apparent that schools are in no position to supply the necessary teaching of them to all of their potential social scientists and natural scientists. Furthermore, if a student has demonstrated at school that he has an aptitude for these subjects, it is unlikely that he will be encouraged to read geography at the university. Against this background, it would seem that such statistical and mathematical understanding as is required might best be fitted into the university syllabus. Here, too, there are problems. Perhaps two streams of teaching might be contemplated—an elementary stream dealt with in departments of geography and an advanced stream provided by departments of statistics and mathematics. One point is clear. If statistics and mathematics are added to the list of ancillary accomplishments for geography graduates, some branches of the subject will have to be rejected to make room for them.

The introduction of quantitative methods to geography poses more fundamental problems. It is unlikely that more than a small proportion of geographers will be competent to handle them. Nor are more than a handful likely to achieve results in geographical research that will make an impact on their mathematical and statistical colleagues. It is not inconceivable that at some time in the future departments of geography will require a professional mathematician or statistician in their own right—unless graduates in these subjects can be persuaded to transfer their allegience to geography at the research level. There are good examples of mathematicians who have become economists—Lord Keynes was one of them.

Whether statistics or mathematics are urged, it is evident that there are large numbers of recruits to present-day geography who lack the facility to handle them but who are respectably literate. It need not be assumed that all future recruits to departments of geography will wish to devote themselves primarily to a quantitative line. It may be in the best interest of most that they should, but the choice should be left to them. The best solution, at least in the short term, is to offer students a variety of courses from which to choose and at the same time to advise them how their combination of choices is likely to open and close doors in the future. Clearly, for most postgraduate work and in certain specific areas of employment such as planning, the new approach with its associated skills is critical. But most geographers will not undertake research either in the subject or in another field when they enter employment. As long as it is realized that employment opportunity and salary scale may be conditioned by the particular kind of geographical training, that is enough. It is to be hoped that all students will not enter university with this as their only motive. There will always be openings for the geographer whose inclinations are towards literacy rather than numeracy.

Whatever choice is made by students, it will speedily become apparent that the quantitative approach is the fashionable approach. At the same time it will become evident that with its feet on the ground of quantitative studies, academic geography is as competent as any of the social sciences to concern itself with prediction. In the past, prediction was regarded as lying outside the field of geographers. Geography, it was argued, was concerned with reality. It looked to the detail of the present and had no concern with the future. A generation ago, geographers would not willingly have subscribed to the motto *Savoir pour prévoir* and such a concept as theoretical geography would have been regarded by them as a contradiction in terms. Yet, even at that time, some geographers were already casting horoscopes. An example is provided by the remarkably accurate prediction made by W. William-Olssen (1941), against the background of statistical trends, for the growth of Stockholm.

Prediction has as its object more than the exercise of the intellect. Geographers *qua* geographers were slow to use their talents in an applied form. They engaged in the collection of data, its representation and analysis, but originally left its application to others. They were familiar with the exigencies of space, aware of the economics of resources, and sensitive to the value of amenities. But their studies tended to have a static rather than a dynamic emphasis, a descriptive rather than an analytical quality. They had an almost innate awareness of distributions on the face of the land, and some of them were keenly aware of the changing patterns through time. What was needed was a fuller recognition of the processes whereby the distributions came into being, of analogues which might help to explain the processes, and of the consequent bases to project trends and tendencies into future situations. In a relatively short time, a small number of geographers, alert to the work of colleagues in kindred disciplines, have begun to assemble a corpus of precedent, procedure, theory, and hypothesis, which is capable of application to a growing range of problems and situations.

Applied geography, worked out in the context of the new approach to the subject, can be an exciting discipline, both on the score of the background training and of the imagination that it demands. Society is in need of people with such interests in its private as well as its public sectors. As an applied study geography has a distinct contribution to make to management. Its concern is with the essentials of location. Though not in general executive its rôle is likely to be increasingly strong in those areas of decision-making which concern themselves with aspects of distribution.

It is hardly surprising that an age which is familiar with the experiences in Aldous Huxley's *Doors of Perception* (1954) should concern itself with states of awareness. Geographers have added perception to precision and prediction as a third quality that they seek to cultivate. In both an artistic and scientific sense, there is a move to approach the visual aspects of the subject with keener understanding. The psychological aspects of perception as they have affected the subject matter of geography have been largely ignored. There was a crude appreciation of the importance of map and diagram as propagandist instruments during the years when *Geopolitik* stirred Europe; while pedagogues have always paid due regard to textbook illustration. The impact of shape and colour, like so much else, has been the object of scientific reappraisal. In the field of geography it has given rise to a perceptionist school of inquiry. To understand the processes whereby the image of a geographical area or feature is created is to be partly aware of the causes that lie behind human behaviour. It is all very well to predict against the background of measurement. Images neither conform nor can they be readily made to conform to seemingly measurable reality. To quote the aphorism of George Perkins Marsh, "Sight is a function; seeing, an art". David Lowenthal (1968) has illustrated this aspect of geography through his reappraisal of "The American Scene". David Harvey (forthcoming) has sought to gather together the theory and practice of this rapidly evolving area of geographical study.

From these few examples, it will be clear that the contemporary geographer is faced with a greater volume of factual material to digest than ever before and it multiplies in an almost geometrical progression. His action in the face of this is governed by the principle of least effort. He must economize in time and effort. The mechanical hardware of the new age of technology can help greatly in the processing and storage of this information. A new language of signs and symbols enables him to state his hypotheses and express his results with a minimum of words. If the situation appears formidable, it is important that he should keep it in perspective. It is common to most disciplines. In some respects, it is a situation easier than that in the physical sciences. At least, there is more of a sense of construction in contemporary geographical studies than there is in the physical sciences which have witnessed the crumbling of old-established laws.

The strength of British geography during the last generation has been inseparable from the strength of geography in the schools, and especially in the sixth forms. The university has received geographers from the school and fed back higher-powered geographers to them. The cumulative effect has been critical. It will be interesting to see if Swedish geography—which in recent years has led the subject in the Old World—will maintain its recruitment now that it has interrupted the system by ceasing to teach geography for university entrance. By inculcating an interest in the countryside through the shared experiences of fieldwork, and by revealing the mysteries of the map, the subject acquires an early appeal. In addition, it is a stabilizing force in a shifting universe of disciplines. At the university level, this stability has been closely related to the broad tolerance that has sprung from a group of people whose investigations cover a wide spectrum. The rivalry between the human and physical geographers has been friendly and fruitful. In the university, geography as a discipline can experience rapid changes; in the school, it cannot. For this reason alone, some traditional concepts must not be swept away too rapidly—always assuming that they should be swept away in any case. They are a part of the early life of the subject and to repudiate them is to throw away a heritage that has given geography its present popularity. In so far as popularity is expressed in numbers, some might argue that geography is too inflated in its recruitment and that to reduce a little of its present adipose tissue would be all to the good. Others might argue that, having built up the present solid base of the subject, it is now poised for a take-off and must be more selective in its recruitment. Should numbers be sustained—or trimmed? The issue may decide itself in the context of broader university changes independently of the changing appeal of the subject.

In the last analysis, the course of geographical knowledge is determined by men and women of imagination and enthusiasm. Imagination is stirred and enthusiasm is communicated in many ways, not in one way. In its present critical phase geographers have to restructure the framework of their concepts and methods without reducing the questionable (if sometimes un-questioning) enthusiasm that their subject has inspired in two generations of practitioners. Enthusiasm is easily crushed; imagination easily destroyed. Geography must not become so mechanistic that it loses its soul, must not become so lost in theory that it forgets the practices of the earth, and must not become so exclusive that it can only accept those who speak its private language.

REFERENCES AND FURTHER READING

BIRD, J. L. and MORRISON, A. (1964) Space photography *in* its geographical applications, *Geographical Review*, **54**, 463–86.
HARVEY, D. W. (forthcoming), *Explanation in Geography*, Arnold, London.
HUXLEY, A. (1954) *Doors of Perception*, Chatto, London.
HUXLEY, A. (1963) *Science and Literature*, Chatto, London.
LOWENTHAL, D. (1968) The American scene, *Geographical Review*, **58**, 61–88.
WILLIAM-OLSSEN, W. (1941) *Stockholms framtida utveckling*, Stockholms kommunal-förvaltning, Stockholm.

A wide range of developments in geography are considered in:

COHEN, S. B. (Ed.) (1967) *Problems and Trends in American Geography*, Basic Books, New York.

Two examples of specific developments in approach are found in:

CURRY, L. (1966) Chance and the landscape, in HOUSE, J. W. (Ed.) *Northern Essays in Honour of G. H. J. Daysh*, Oriel P., Newcastle-upon-Tyne, pp. 40–55.
McNEE, R. B. (1960) Towards a more humanistic economic geography, the geography of enterprise, *Tijdschrift voor Economische en Sociale Geografie*, **50**, 201–6.

Examples of the broader contemporary relationships of geography are evident in:

BOULDING, K. E. (1956) *The Image*, U. of Michigan P., Ann Arbor.

POPPER, K. R. (1962) On the logic of the social sciences, *Kölner Zeitschrift für Soziologie und Socialpsychologie*, **14**.

ZIPF, G. K. (1949) *Human Behaviour and the Principle of Least Effort*, Addison-Wesley P., Cambridge, Mass. (A facsimile of this edition was published in 1965 by Hafner New York.)

Some apposite reflections on geography and the university milieu are found in:

WHEATLEY, P. (1968) Great expectations, *The Bloomsbury Geographer*, **1**, 1–9.

PART I

PHYSICAL GEOGRAPHY

In Nature's infinite book of secrecy
A little I can read.

SHAKESPEARE, *Antony and Cleopatra*

CHAPTER 2

PROGRESS IN GEOMORPHOLOGY

R. U. COOKE

INTRODUCTION

Examiners of geomorphology scripts at all educational levels are often reminded of persistent errors of fact, emphasis, and omission which plague the subject. How often it is said, for example, that deltas are formed in tideless seas; that the longitudinal profile of a river valley may be divided into three sections which are respectively young, mature, and old; that terraces are only formed following a fall of sea-level; and that desert landforms are characteristically angular. It would be an impossible task to review and correct such errors here. Some common grumbles of geomorphology examiners have been effectively aired in the pages of *Geography* (Dury, 1963; Pugh, 1964; Small, 1966).

Errors persist, of course, largely because available textbooks are so frequently based upon out-of-date material. Analysis of references in the latest editions of widely used elementary texts—references which are presumably representative of the authors' sources—shows that only a few books, such as Dury's inexpensive *The Face of the Earth* (Pelican, 1966), incorporate a substantial body of data published since 1950. Antediluvian sources would matter little if geomorphology had stood still since 1950, but it has not. The great flood of geomorphological publication is swiftly rising,* and recently there have been rapid and important changes in the directions of geomorphological research. As a result the gulf between teachers and research workers widens. An attempt is made here to bridge that gulf by identifying and briefly reviewing some of the principal changes which have occurred in geomorphology, by commenting on their relevance to the teaching of the subject, and by drawing attention to published works which exemplify the

---

* *Geographical Abstracts, Series A (Geomorphology)* includes brief abstracts of much of this material, and is the best single source of reference for teachers and students who wish to keep up to date. There are three other series which cover biogeography and climatology (B), economic geography (C), and social geography and cartography (D). The periodicals can be purchased from *GeoAbstracts*, University of East Anglia, University Village, Norwich, NOR 88C, England.

changes. Other reviews by Clayton (1964) and Peel (1967) will be of interest to many readers.

THE STUDY OF FORM
Data from the Field

The overwhelming tendency in recent years has been towards more precise, quantitative field measurements of landforms and deposits. For example, after decades of qualitative slope descriptions, slope inclinations are now invariably measured instrumentally; seismic refraction and electrical resistivity equipment is often used to record the thicknesses of superficial deposits; and strength and permeability of soils are measured by a wide variety of techniques.

Qualitative observation of form has now been almost completely superseded by techniques of morphological mapping which have been devised in many parts of the world to allow the recording of topographic detail on maps or, more commonly, on airphotographs. Most of the techniques aim either to subdivide the landscape into areas of approximately uniform properties which are separated from adjacent areas by boundaries recognizable in the field, or to delimit specific landform types. Figure 1 shows part of a morphological map of Eskdale, England, compiled using a successful technique developed at the University of Sheffield (Waters, 1958; Savigear, 1965), in which discrete units of fairly uniform slope inclination are delimited by either breaks or changes of slope.

Many properties of these units can be measured precisely, including, for example, geometric properties (relief, area, shape, slope, orientation, etc.), and properties of features on, in, and under the surface (lithology of bedrock, permeability of soil, density of vegetation, etc.). These data are of great value in studying the nature of a landscape and its constituent parts. An analysis of such data, and a discussion of the possibilities and pitfalls associated with it, is presented by Gregory and Brown (1966) who investigated the slope angles of units to see how they varied areally with such factors as geology and orientation.

Data from Maps

There has been a pronounced movement away from description based upon simple visual inspection of maps, characterized by vague and ill-defined phrases like "dendritic drainage patterns" and "gently undulating topography", towards precise techniques for the recovery and analysis of morphometric data presented on maps (Clarke, 1966; Slaymaker, 1966; Zakrzewska, 1967). Many recovery techniques are straightforward and are exemplified by

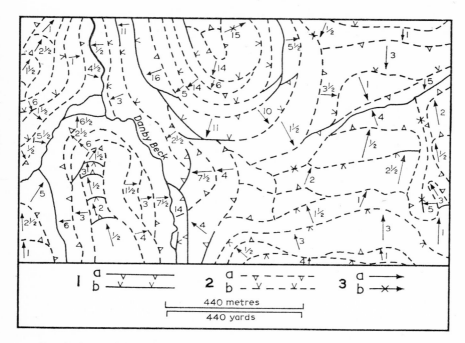

Fig. 1. A morphological map of part of Eskdale, England. (After Gregory and Brown, 1966.) 1, break of slope: (a) convex (b) concave. 2, change of slope: (a) convex (b) concave. 3, direction and angle of slope in degrees: (a) rectilinear (b) convex.

devices for measuring slope (Thrower and Cooke, 1968) and shape (Stoddart, 1965).

Data collected by these methods may provide valuable insights into erosional history, and the present relations between features in drainage basins, when they are analysed using various graphical and quantitative* techniques.

Rewards from morphometric analysis include:

(1) The revelation of organization in the spatial distribution of landforms. For example, analysis of the length and numbers of streams and the areas of

* Quantitative methods widely used at present are mostly statistical techniques which are outlined in many inexpensive books, including Reichmann, W. J. (1962) *The Use and Abuse of Statistics*, Pelican. Harmondsworth, and Hayslett, H. T. (1968) *Statistics Made Simple*, Allen, London. The use of quantitative methods has been reviewed and vigorously advocated by R. J. Chorley in Chorley R. J. and Haggett, P. (Eds.) (1966) *Frontiers in Geographical Teaching*, Methuen. London, pp. 147–63, and in Dury, G. H. (Ed.) (1966) *Essays in Geomorphology*, Heinemann, London, pp. 215–387. The innovation of quantitative methods of analysis is closely associated with the recent development of data-processing machinery and techniques for acquiring quickly much more data. For comments on the use of quantitative methods in human geography, see Chapter 8, below.

drainage basins shows that these properties are usually related in a simple and generally predictable way both to stream order* and to one another. Indeed, these and other similar statistical relations appear sufficiently well-established to be described as "laws of drainage composition" (Morisawa, 1968). Although such analysis is most appropriate to the study of drainage basins, it may also be applicable to the study of karst, and other landforms (Slaymaker, 1966).

(2) The provision of a powerful means of regional comparison. Chorley and Morgan (1962) determined the average values of various morphometric properties in order to construct two idealized drainage basins representing the Unaka Mountains in Tennessee and North Carolina, and the Dartmoor Uplands, England. They demonstrated convincing differences of drainage density, stream length, drainage basin area, and relief between the two areas.

(3) The evocation of different interpretative problems. In the previous illustration, for instance, one might ask "why do the differences exist?". Chorley and Morgan argued that many of the differences may be explained on the basis of consistently differing rainfall intensities received by the two areas since the Miocene period.

(4) The testing of some old, qualitative generalizations. In the American Southwest it has been suggested that pediment slopes are inversely related to drainage catchment areas of pediments because, deductively, it would appear likely that a larger drainage catchment will discharge more water, which in turn would develop a lower pediment slope. Measurement of these two morphometric properties in the Mojave Desert, California, and analysis of the relations between them by the author, showed that they are not significantly related; other explanations of pediment slope had to be explored.

(5) The partial explanation of forms. For example, McConnell's (1966) analysis of data, which was derived largely from maps, revealed that topographic slope on various dissected glacial tills of the Upper Mississippi Valley is principally a function of dissection, position, orientation, age and type of materials, and especially available relief.

Geomorphological Description

As landform description becomes more precise so the need declines for figurative and anthropomorphic terms (Dury, 1963). It is no longer necessary to represent landscapes in the hackneyed and often inaccurate terms of "youth", "maturity", and "old age", when simple objective terminology

* Several schemes of stream ordering have been developed, but a commonly used system defines *first-order stream segments* as those without tributaries, *second-order stream segments* as those segments of streams produced by the joining of two first-order stream segments, etc. The study of this topic has been developing rapidly in recent years.

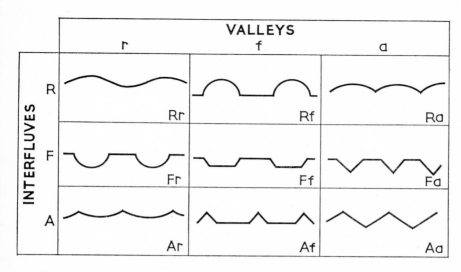

Fig. 2. Classification of landscape profiles based on interfluve profiles and valley profiles. (After Ollier, 1967.)

will suffice. Figure 2 shows a classification of landscape profiles based upon the association of interfluve and valley profiles which may be rounded, angular, or flat (Ollier, 1967). (The nomenclature used in this diagram is, of course, only a form of shorthand, not a new and mysterious jargon.) As Ollier (1967, p. 78) explained, the profiles in Fig. 2 might be described using traditional stage names as follows:

Profile	Valley	Landscape
Rr	Mature?	Mature
Rf	Mature/old	Mature
Ra	Young	Mature
Fr	Mature?	Young
Ff	Mature	Young
Fa	Young	Young
Ar	Mature?	Late mature
Af	Mature/old	Late mature
Aa	Young	Mature

It is clear from this list that the description of landscapes by means of stage names is ambiguous and imprecise. For example, six valley forms are possibly "mature" and three valleys are "young"; the "mature" landscape Aa is composed entirely of "young" valleys! A most welcome and profitable change in landform teaching might be made with the use of simple descriptive techniques such as that developed by Ollier.

The virtues of precision and quantification in geomorphological description are manifest. One's observations can be analysed, repeated, tested, extended, and understood by others. Few would contend that description is an end in itself, but it is always necessary, and unless it is precise it may be of little further value. Measurement focuses attention on the geographer's fundamental problems of scale and areal comparison. Durable relationships between forms, and between forms and processes, may be established. Attention is focused on potentially soluble problems and geomorphology is firmly transplanted into the body of natural science.

THE STUDY OF PROCESS

The nature and rates of landform development processes, for so long assumed or disregarded by many geomorphologists, are now being studied in the field and simulated in the laboratory by a small but growing number of research workers.

Processes in the Field

In some studies, such as those of coastal landforms, observation of processes has been of great importance for many years. In other studies, research has been revitalized by the renewed recognition that a knowledge of current processes is indispensable for realistic interpretations of both present and past landscapes. For example, detailed field investigations of present activity of glacier ice (e.g. Price, 1966; Hewitt, 1967) are of the greatest value in explaining the origins of landforms in formerly glaciated landscapes.

Experimental measurement of processes in the field is another fashionable trend. Rates of soil creep, for instance, have now been studied by many different techniques ranging in sophistication from the measurement of tree curvature to the use of fairly elaborate experimental designs (e.g. Kirkby, 1967). There are many examples of enlightening results gained from process studies which employ simple and inexpensive equipment. For example, a periglacial area is usually said to be dominated by freeze–thaw action and solifluction, but a study by Rapp (1960) in northern Scandinavia showed in that area almost as much material is being removed by solution as by all other processes combined, and solifluction is of relatively little significance.

From the teaching viewpoint, it is much easier and cheaper to undertake fairly accurate studies of processes in the field than might be imagined. Most schools are in a position to adopt sites which are rapidly changing—such as quarries, roadcuts, tip heaps, and erosional scars—and much can be done by the seaside (Sparks, 1949) or in the channels of small streams. Significant changes can often be observed and important general principles effectively learnt at such sites. For instance, the rate and nature of channel bank development may be recorded by driving steel pins (or long nails) into the banks until they are flush with the surface, and then measuring at regular intervals the extent to which the pins become exposed or buried (e.g. Emmett, 1965). Many other simple field techniques are described in the *Revue de Géomorphologie Dynamique*, part 4 (1967).

Laboratory Simulation of Processes

Experimental studies of processes in the laboratory are becoming much more common. Mention may be made of investigations of frost weathering (Wiman, 1963), ice movement (Lliboutry, 1965), and alluvial fan formation (Hooke, 1967), and of numerous flume studies of fluvial processes (e.g. Guy *et al.*, 1966). Results from many of these controlled process studies are fundamental to our understanding of the real world.

Simple classroom experiments can demonstrate splendidly the nature of many geomorphological processes. For example, features developed in proximity to stagnant ice can be reproduced on an inexpensive stream-table (Schwartz, 1963); and a simple but effective flume, an outstanding teaching device for the study of water and sediment movement, can be constructed by an ingenious teacher with a little capital (Harrison, 1967).

EXPLANATION OF LANDFORMS

Dynamic Geomorphology

There have been considerable advances in recent years in the examination of forces (e.g. running water, gravity, etc.) acting on resistances (e.g. rocks, etc.) to produce landforms which may be in or approaching a condition of equilibrium (Strahler, 1952). It is acknowledged that most landscapes are the result of activity within extremely complex working systems incorporating many forces and resistances which interact with one another, so that a change in any one variable usually affects many of the others. To study the whole of a working system is an exceptionally difficult, if not impossible, exercise, and the research worker is commonly forced to select parts of a system for analysis. Three practical expedients may be mentioned:

(1) Association of two variables. It can be demonstrated, for example, that channel depth and channel width are related closely, but not perfectly, to discharge of water in a channel (e.g. Leopold and Miller, 1956).

(2) Association of more than two variables. In an outstanding contribution, Melton (1957) investigated quantitatively the relations between carefully measured elements of climate, surface properties, and landforms. He demonstrated, for example, that drainage density (the total length of drainage lines per unit area) in selected areas of the western United States may be largely predicted in terms of a precipitation effectiveness index, the percentage of bare ground, soil strength, and infiltration capacity, and run-off intensity, and he established precisely the relative importance of these variables.

(3) Analysis of the net effect of many variables when only some are known and measured. In view of the apparent complexity of natural working systems, and the extreme difficulty of measuring all the variables within them, Scheidegger and Langbein (1966) have suggested that the net effect of all the processes acting simultaneously or sequentially can be studied by assuming that the processes occur at random, and by predicting the probable results.* Slope profile development, for instance, has been studied in this way. The use of probability theory is but one recent example of many attempts which have been made to analyse landforms and processes in terms of theoretical and deductive models.

Because of difficulties of collecting and analysing data, there are at present few studies in the field of dynamic geomorphology. But these studies have been successful in establishing meaningful and precise relations between form and process. In addition, such studies pave the way towards practical application of geomorphology to problems of economic significance such as flooding and erosion. And the recognition of man's place in many geomorphological systems promises to provide studies of greater direct value and interest to geographers. For these reasons, studies in this field represent a trend which is likely to become much more important in the future. Complex working systems are more difficult to teach than evolutionary schemes, for the mind is more capable of comprehending consecutive events than grasping the complex interrelations among a large number of contemporaneous events. In addition a knowledge of quantitative methods is desirable. But important contemporary relations between variables can be demonstrated diagrammatically (e.g. Chorley, 1967, p. 80), and discussed by separate consideration of the variables and the links between them.

Climate and Geomorphology

Two profitable approaches to the examination of landforms are *climatic geomorphology*, "studies of contemporary or fossil climatic influence on land-

* These results are based on the use of probability theory which provides precise methods of determining probable solutions to problems by the analysis of incomplete data and by taking account of chance effects. Probability theory is simply described in Reichmann, W. J. (1962) *The Use and Abuse of Statistics*, Pelican, Harmondsworth, chapter 14.

forms", and *climatogenetic geomorphology*, "the study of different generations of surface forms within a single landscape which have resulted from both contemporary and fossil climatic influences" (Holzner and Weaver, 1965, p. 592). Both owe much to the early views of W. M. Davis, V. V. Dokuchayev, and Albrecht Penck, and they have been favoured and developed for several years in Europe (e.g. Büdel, 1959; Birot, 1968). The principal recent advances in these fields include the identification and study of morphogenetic regions (e.g. Birot, 1968); more precise evaluation of relations between climate and process, and process and form at present and in the past (e.g. Schumm, 1965); and a fuller appreciation of the nature and importance of climatic change (e.g. Dury, 1967, and Fig. 3). There is a marked dichotomy between these approaches, which recognize regional associations between climate and landform, and the view of certain geomorphologists, epitomized by L. C. King's monumental *The Morphology of the Earth* (1967), that many important processes operate in most environments, differ only in degree from place to place, and produce essentially similar forms.

There is considerable confusion at present in the intractable literature concerned with the relations between climate and geomorphology: the importance of bedrock lithology and structure is often neglected, and most classificatory schemes are open to serious criticisms of overgeneralization. Nevertheless, it is possible that the relations between climate and landforms may yet provide a useful approach to the teaching of geomorphology especially because they also provide valuable links with other branches of physical geography, notably hydrology and biogeography.

Regional–Chronological Studies

Interest in regional–chronological studies continues unabated, and publications in this field are still a major feature of the geomorphological literature. As methods of studying process and form become more refined, so regional work becomes more intensive, and more concerned with smaller areas. Chronologies of landform development have been greatly improved by the extensive use of sedimentary records in addition to erosional evidence, and by the application of recent absolute and relative dating techniques. For example, at least ten geochemical methods have been developed recently for determining the absolute age of Quaternary events (Broecker, 1965). Brief review of this enormous field is impossible here: a single study effectively exemplifies the literature which incorporates these changes. Kerney *et al.* (1964) investigated a small section of the chalk escarpment near Brook (Kent) containing seven steep-sided dry valleys with which are associated deposits of chalk debris, filling their bottoms and extending as lobes over the Gault Clay plain beyond. Evidence from detailed morphological and stratigraphical

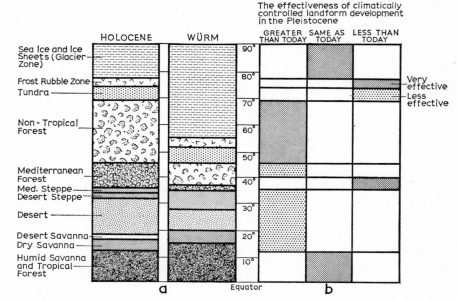

Fɪɢ. 3. Climatic change and landform development in the Quaternary Era: a major
focus of research interest.
(a) Differences in the latitudinal distribution of climatic zones in the northern
hemisphere between 10°W and 35°E during the last cold phase (Würm) and the
present warm phase (Holocene) of the Quaternary Era. The complexity of these
relations is compounded several times when the whole of Quaternary time is
considered, for there were at least six cold periods separated by warmer periods.
(b) A greatly generalized picture of the relative importance of climatically-controlled
landform development in the Pleistocene and today in the same area shown in
Fig. 3a. Where Pleistocene processes were powerful and processes are relatively weak
today, fossil landforms may still be preserved in the landscape (left-hand column);
where processes are strong today, Pleistocene landforms may have been transformed
into landforms in harmony with present conditions (right-hand column); and where
there has been little change in either the nature or intensity of processes, landforms
may have been developed continuously throughout the Quaternary (central
column). (After Büdel, 1959.)

mapping, isotope dating, and analysis of pollen, mollusca, and artifacts led
to the construction of an unequivocal and comprehensive history of the area
which adds significantly to our knowledge of dry valley development in
southeast England.

For many years, geomorphology teaching has emphasized landscape
evolution, and this emphasis will probably persist for some time; but as
detailed knowledge of landscape development slowly progresses, it becomes
clear that general and highly controversial evolutionary schemes, such as
those of W. M. Davis and W. Penck, are inadequate and unproven. One

must seriously doubt that it is either possible or desirable to teach immature students a comprehensive story of landscape evolution when research workers are themselves uncertain what the story is. Perhaps the schoolteacher should abandon these broad schemes and concentrate on explaining detailed and well-documented evolutionary sequences of different areas at various scales. There are several sources which furnish the teacher with suitable studies undertaken in many areas: *A Bibliography of British Geomorphology* (Clayton, 1964); *The Quaternary of the United States* (Wright and Frey, 1965); L. C. King's *The Morphology of the Earth* (1967), and *Landform Studies from Australia and New Guinea* (Jennings and Mabbutt, 1967).

CONCLUSION

Changing methodology and changing techniques have greatly altered the study of geomorphology. Some of the changes have been accompanied by verbal conflicts which often reflect fashion and differences of emphasis. For example, in an important field of debate, those preoccupied by historical geomorphology who inherit essentially cyclic views of landscape development are clearly at odds with those who deliberately neglect the time dimension and practise an essentially non-cyclic approach to landform study. Fashion is here reflected in the curious contradiction that at a time when some geomorphologists have been moving from cyclic to non-cyclic interpretations of landforms (e.g. Hack, 1960 and 1965; Chorley, 1962), some pedologists—whose familiarity with landforms is as great as that of many geomorphologists—have been moving in precisely the opposite direction in their interpretation of soils (e.g. Butler, 1959). The differences of emphasis between cyclic and non-cyclic views are not as great as some have maintained, and rational compromise between them is possible (e.g. Schumm and Lichty, 1965). This and other controversies will, of course, continue to provide welcome stimulation to the subject; but there are still too many sides to every mountain, and too much is hidden in each stream channel for any single approach to be acceptable to all.

Some more permanent, less controversial and long overdue changes include the introduction of precision, attention to detail, experimentation, the use of quantitative methods and the pruning of some unacceptable old material and ideas. These changes are fundamental and will eventually penetrate geomorphology at all its levels: already much can usefully be done by introducing field and laboratory experiments, morphological mapping, and morphometric analysis exercises, and especially by encouraging field observation. When the time comes to rewrite geomorphology syllabuses and incorporate these changes, a strong case can be made for a simple uniformitarian approach which begins with first principles (such as the strength of rock and the nature of water flow) and makes a minimum of basic assumptions. Such an approach would include methods of precise description of landforms, deposits, and processes, and would attempt to evaluate the relations between these features in various areas. Building on such a sure foundation, the student can then go forward to study problems in applied geomorphology and also begin to penetrate the labyrinth of evolutionary concepts with open eyes. Some teachers may already follow this pattern, but a general change can come only with new textbooks and syllabuses, for we are all caught in that most inflexible cycle of progressive degradation which begins with the author and passes through the stages of publisher, teacher, student, and examiner.

REFERENCES AND FURTHER READING

INTRODUCTION

Some geomorphological grumbles are aired in:

DURY, G. H. (1963) Rivers in geographical teaching, *Geography*, **48,** 18–30.
PUGH, J. C. (1964) Some avoidable errors in physiographic studies, *Geography*, **49,** 44–9.
SMALL, R. J. (1966) Some criticisms of the teaching of geomorphology at A-level, *Geography*, **51,** 29–37.

Reviews of trends in geomorphology include:

CLAYTON, K. M. (1964) A face-lift for landform studies?, *New Scientist*, **24,** 103–6.
PEEL, R. F. (1967) Geomorphology: trends and problems, *Advancement of Science*, **23,** 128–65.

THE STUDY OF FORM

Progress in the collection of field data is exemplified by:

GREGORY, K. J. and BROWN, E. H. (1966) Data processing and the study of land form *Annals of Geomorphology*, **10,** 237–63.
SAVIGEAR, R. A. G. (1965) A technique of morphological mapping, *Annals of the Association of American Geographers*, **55,** 514–38.
WATERS, R. S. (1958) Morphological mapping, *Geography*, **43,** 10–17.

Recovery of geomorphological data from maps is considered in:

CLARKE, J. I. (1966) Morphometry from maps, in DURY, G.H. (Ed.), *Essays in Geomorphology*, Heinemann, London, pp. 235–74.
CHORLEY, R. J. and MORGAN, M. A. (1962) Comparison of morphometric features, Unaka Mountains, Tennessee and North Carolina, and Dartmoor, England, *Geological Society of America Bulletin*, **73,** 17–34.
McCONNELL, H. (1966) A statistical analysis of spatial variability of mean topographic slope of stream-dissected glacial materials, *Annals of the Association of American Geographers*, **56,** 712–28.
MORISAWA, M. (1968) *Streams—their Dynamics and Morphology*, McGraw-Hill, New York.
SLAYMAKER, H. O. (Ed.) (1966) Morphometric analysis of maps, *British Geomorphological Research Group Occasional Paper*, **4.**
STODDART, D. R. (1965) The shape of atolls, *Marine Geology*, **3,** 369–83.
THROWER, N. J. W. and COOKE, R. U. (1968) Scales for determining slope from topographic maps, *Professional Geographer*, **20,** 181–6.

Zakrzewska, B. (1967) Trends and methods in land form geography, *Annals of the Association of American Geographers*, **57**, 128–65.

Two valuable commentaries on geomorphological description are:

Dury, G. H. (1963) Geographical description: an essay in criticism, *Australian Geographer*, **9**, 67–78.
Ollier, C. D. (1967) Landform description without stage names, *Australian Geographical Studies*, **5**, 73–80.

THE STUDY OF PROCESS

Studies of processes in the field are exemplified by:

Emmett, W. W. (1965) The VIGIL network: methods of measurement and a sampling of data collected, *International Association of Scientific Hydrology Publication*, **66**, 89–106.
Hewitt, K. (1967) Ice-front deposition and the seasonal effect: a Himalayan example, *Transactions of the Institute of British Geographers*, **42**, 93–106.
Kirkby, M. J. (1967) Measurement and theory of soil creep, *Journal of Geology*, **75**, 359–78.
Price, R. J. (1966) Eskers near the Casement Glacier, Alaska, *Geografiska Annaler*, **48**, 111–25.
Rapp, A. (1960) Recent developments of mountain slopes in Kärkevagge and surroundings, northern Scandinavia, *Geografiska Annaler*, **42**, 65–206.
Sparks, B. W. (1949) Geomorphology by the seaside, *Geography*, **34**, 216–20.

Laboratory simulations of geomorphological processes are exemplified by:

Guy, H. P., Simons, D. B. and Richardson, E. V. (1966) Summary of alluvial channel data from flume experiments, 1956–1961, *U.S. Geological Survey Professional Paper*, **462-I**.
Harrison, S. S. (1967) Low-cost flume construction, *Journal of Geological Education*, **15**, 105–8.
Hooke, R. Le B. (1967) Processes on arid-region alluvial fans, *Journal of Geology*, **75**, 438–60.
Lliboutry, L. (1965) How glaciers move, *New Scientist*, **28**, 734–6.
Schwartz, M. (1963) Stream-table development of glacial landforms, *Journal of Geological Education*, **11**, 29–30.
Wiman, S. (1963) A preliminary study of experimental frost weathering, *Geografiska Annaler*, **45**, 113–21.

EXPLANATION OF LANDFORMS

Contributions to dynamic geomorphology include:

Chorley, R. J. (1967) Models in geomorphology, in Chorley, R. J. and Haggett, P. (Eds.) *Models in Geography*, Methuen, London, pp. 59–96.
Leopold, L. B. and Miller, J. P. (1956) Ephemeral streams—hydraulic factors and their relation to the drainage net, *U.S. Geological Survey Professional Paper*, **282-A**.
Melton, M. A. (1957) An analysis of the relations among elements of climate, surface properties, and geomorphology, *Columbia University Department of Geology Technical Report*, **11**.
Scheidegger, A. E. and Langbein, W. B. (1966) Probability concepts in geomorphology, *U.S. Geological Survey Professional Paper*, **500-C**.
Strahler, A. N. (1952) The dynamic basis of geomorphology, *Geological Society of America Bulletin*, **63**, 923–8.

The relations between climate and geomorphology are considered in:

Birot, P. (1968) *The Cycle of Erosion in Different Climates*, Batsford, London.
Büdel, J. (1959) The "periglacial"-morphologic effects of the Pleistocene climate over the entire world, *International Geology Review*, **1**, 1–16.
Dury, G. H. (1967) Climatic change as a geographical backdrop, *The Australian Geographer*, **10**, 231–42.

HOLZNER, L. and WEAVER, G. D. (1965) Geographic evaluation of climatic and climato-genetic geomorphology, *Annals of the Association of American Geographers*, **55,** 592–602.

KING, L. C. (1967) *Morphology of the Earth*, 2nd edn., Oliver & Boyd, Edinburgh.

SCHUMM, S. A. (1965) Quaternary paleohydrology, in WRIGHT, H. E., JR., and FREY, D. G. (Eds.) *The Quaternary of the United States*, Princeton U. P., Princeton, pp. 783–94.

The enormous literature in the field of regional–chronological studies is exemplified by:

BROEKER, W. S. (1965) Isotope geochemistry and the Pleistocene climatic record, in WRIGHT, H. E., JR., and FREY, D. G. (Eds.) *The Quaternary of the United States*, Princeton U.P., Princeton, pp. 737–53.

CLAYTON, K. M. (Ed.) (1964) A Bibliography of British Geomorphology, *British Geomorphological Research Group Occasional Paper*, **1.**

JENNINGS, J. N. and MABBUTT, J. A. (Eds.) (1967) *Landform Studies from Australia and New Guinea*, Cambridge U.P., Cambridge.

KERNEY, M. P., BROWN, E. H. and CHANDLER, T. J. (1964) The Late-Glacial and Post-Glacial history of the chalk escarpment near Brook, Kent, *Philosophical Transactions of the Royal Society of London*, series B, **248,** 135–204.

WRIGHT, H. E., JR., and FREY, D. G. (Eds.) (1965) *The Quaternary of the United States*, Princeton U.P., Princeton.

CONCLUSION

The rôle of time in landscape development is considered in:

BUTLER, B. E. (1959) Periodic phenomena in landscapes as a basis for soil studies, *C.S.I.R.O. Australian Soil Publication*, **14.**

CHORLEY, R. J. (1962) Geomorphology and general system theory, *U.S. Geological Survey Professional Paper*, **500-B.**

HACK, J. T. (1960) Interpretation of erosional topography in humid temperature regions, *American Journal of Science*, **258-A,** 80–97.

HACK, J. T. (1965) Geomorphology of the Shenandoah Valley, Virginia and West Virginia, and origin of the residual ore deposits, *U.S. Geological Survey Professional Paper*, **484.**

SCHUMM, S. A. and LICHTY, R. W. (1965) Time, space and causality in geomorphology, *American Journal of Science*, **263,** 110–19.

In addition to the books mentioned above, the following reference works are of interest:

BUTZER, K. W. (1964) *Environment and Archaeology*, Methuen, London.

KING, C. A. M. (1966) *Techniques in Geomorphology*, Arnold, London.

LEOPOLD, L. B., WOLMAN, G. M. and MILLER, J. P. (1964) *Fluvial Processes in Geomorphology*, Freeman, San Francisco.

CHAPTER 3

RECENT ADVANCES IN METEOROLOGY AND CLIMATOLOGY AND THEIR RELEVANCE TO SCHOOLS

T. J. CHANDLER

THE STATUS OF METEOROLOGY AND CLIMATOLOGY

In many schools and perhaps even some universities, meteorology and climatology are clearly regarded as something of a geographical *bête noire* by both teachers and taught. Some teachers seem to have almost capitulated in the face of some real and many imaginary difficulties and, by way of compensation, have developed the art of circumnavigating the problems by concentrating on alternative topics which are frequently optional in geography syllabuses. This is both unfortunate and unwarranted.

There is little doubt that in a majority of schools, standards of attainment in meteorology and climatology are generally well below those in most other parts of the geography syllabus. And yet the majority of students and teachers who shy away from these subjects, apparently thinking them either scientifically too difficult or too heavily weighted with data, would be the first to admit their basic relevance to geography. The shape of the ground, its vegetation cover, and the use man makes of the land, topics central to school geography, are undeniably related to climatic controls.

The case for including meteorology and climatology (the latter depending upon the former) in the geography syllabus is irrefutable and few would wish to see them excluded. Their place is warranted both for their own sake as part of the sensible environment and landscape and as essential ingredients of the whole geographical complex. An awareness and understanding, however rudimentary, of the atmospheric environment, is clearly a necessary part of any general geographical training.

Having accepted the need for a meteorological and climatological element in geography, little progress will be made unless the inherent difficulties of the subjects are recognized. Meteorology and its temporal and spatial integrater, climatology, are precise and, in many respects, complex sciences. We cannot close our eyes to the problems consequent upon this; to do so would do geography harm and rightly discredit it and us in the eyes of others. One

27

might, of course, argue that geographers should content themselves with meteorological and climatological descriptions without delving too deeply into questions of genesis, that is with the answers to the question "what?" rather than "why?". This attitude, sometimes also advocated for the study of landforms, need not concern us too much for most geographers have strongly rejected this sterile approach.

A major difficulty experienced in both schools and universities is to keep abreast of advances in meteorology and climatology. This is not easy and it would be foolish to think it was for more knowledge and understanding of the atmosphere have been accrued since the Second World War than in all previous history. But in this respect, meteorology and climatology are by no means unique. The position is the same in most sciences and many arts subjects and the speed of advance is, in any case, no greater than the rapidity with which new books appear on the bookshelves. Fortunately, up-to-date publications are available in meteorology and climatology (Barry, 1967; Barry and Chorley, 1968; Pedgley, 1962; Riehl, 1965) with descriptions and explanations simply, though still scientifically, expressed. The facts are not difficult to obtain in a form suitable for schools.

Perhaps a more relevant difficulty is the fact that relatively little meteorological research has been the work of geographers despite their important contributions to climatology and boundary-layer studies. The great bulk of post-war meteorological research has been by physicists and mathematicians whose publications others may find difficult to understand, frequently because they lack any knowledge of basic physics, which is absolutely vital; without it, meteorology and climatology are virtually impossible. Meteorology without too much mathematics is clearly feasible; meteorology without physics is a contradiction in terms.

In schools, the necessary physics can often be provided as part of a wider training in science. Meteorological processes such as evaporation, condensation, and precipitation can be integrated very successfully into a course in heat physics, and wind-force balances can be taught as examples of simple mechanics. Such training in basic principles should help to remove the widespread confusion amongst geography students between condensation and precipitation processes and such misconceptions as the notion that winds blow from high to low pressure.

Whilst recognizing that it is not only possible but frequently, though not always, justifiable to teach meteorology without mathematics (but not without simple physics), it is very undesirable to teach meteorology and climatology without numbers. A primary aim of sixth-form teaching in meteorology and climatology should be to give the pupils a reasonably detailed quantitative knowledge of the world's climates. This, it is suggested, should be based upon a sound "numerical experience" of climate.

Figures alone mean little unless they can be positioned in a framework of personal experience and the climate of most places over as little as two or three years usually provides short-period illustrations of a very wide range of climatic experiences. Practical measurement and observation, ideally linked to weather-map analysis (Taylor and Yates, 1967), can usefully transform climatological abstraction into meaningful reality. Rainfall rates in the humid tropics and winter temperatures in central North America mean much more if pupils have a personal, numerical experience of a variety of weather. It is obvious that the whole range of world conditions cannot be simulated at one place, but this is not the primary aim. What is important is that well-recognized points of scale reference are established, so that others elsewhere in the world can be realistically related to them. This can be achieved by maintaining a school weather station (Bayliss, 1962), but it is very important that its results are used and seen to be useful. For the more ambitious, soil temperatures measured in artificial soil pits can teach the relevance of soil conditions upon climate and grass (or some other crop) growth very much more effectively than any learnt statement. Teachers will think of many such simple experiments, the instruments for which can often be made at school (Hookey, 1968). Many atmospheric processes can also be taught by simple physical analogies which demand little apparatus but make a point far more forcefully than the written or spoken word. For instance, the "cloud" in the neck of a lemonade bottle when it is first opened is a simple illustration of condensation consequent upon falling pressure.

There remains the difficulty of defining the scope and content of meteorology and climatology syllabuses. Current syllabuses are disturbingly similar at all levels. We therefore need to define the scope and content of the subjects more clearly at different levels in the light of recent advances, and nowhere is this need greater than at sixth-form level (or its equivalent). The answer is, it is suggested, to teach climatic analysis in appropriate synoptic terms, that is in terms of general circulation patterns, airstream characteristics, and meso-scale weather systems, rather than the now outmoded, static, numerical approach in terms of often meaningless, single-element averages. The detail and scope of treatment can then be adjusted to the level of study by varying the rigour of the analyses whilst retaining the continuity of approach. In this respect the general circulation of the atmosphere, whose true nature has emerged only in the last two or three decades and largely as the outcome of post-war technological advances, provides a most satisfactory unifying framework for the description of the world's climates. It enables us to escape from the static, sterile approach of pre-war climatology which was so boring to students who were made to learn, almost by rote, a list of climatological capes and bays. Phenomena as divergent in scale as prevailing ocean currents and seasonal winds on the one hand and thunderstorm overturnings on the other, when integrated into a study of the general circulation of the atmosphere, can all be seen as part of a unified zonal, meridional, and vertical mechanism of energy transfer to restore local balances.

A SUGGESTED SIXTH-FORM SYLLABUS

In the light of these very brief, general considerations, and bearing in mind recent trends and discoveries, training in meteorology and climatology of sixth-form students or their equivalents elsewhere might be organized into four sections as follows:

(1) A short course, perhaps developing an earlier elementary course, in basic meteorological processes. It would be concerned with such factors as evaporation, condensation, precipitation, and force balances upon winds, and could form part of a broader science training and be related to the analysis of readings from a school (or other nearby) weather station.

(2) A course in the energy and water exchanges and balances of the earth–atmosphere system. The course would be in simple terms and would have as a principal aim an understanding of the primary controls upon world climates. The study, in addition to its consideration of energy and water transport would include such features as convergence and divergence, uplift and subsidence and the relation of these to the development of synoptic systems such as depressions, anticyclones, fronts, etc. From this, the general pattern of world climates would be derived.

(3) The more detailed synoptic climatology of chosen areas.

(4) A study of local climates, namely the impacts of the physical and biotic conditions and man's actions, both urban and rural, upon conditions in the atmosphere's boundary layer. This could form a foundation for regional studies of agriculture, etc. Recent advances in each of these four sections are briefly reviewed below.

Recent Advances in Meteorology

Recent advances in our understanding of basic meteorological processes, the first of the four sections in the syllabus mentioned above, have covered a wide variety of individual topics and in most cases have added detail to the already well-known outlines of physical processes. Much of the research has been concerned with various aspects of cloud physics and many of the pre-war meteorological enigmas, of raindrop and hailstone formation for instance, have been resolved, though much important detail remains to be added (Mason, 1962; Pedgley, 1962).

Recent Advances in Climatology

It is within the fields of the general circulation and the energy and water balances of the earth–atmosphere system, the second section of the suggested syllabus outlined above, that some of the greatest advances have been made into what were previously very poorly or even completely misunderstood branches of the subject. One consequence, quite apart from the improved understanding of individual phenomena, has been the integration of these in a system of heat transfer organized to restore local momentum and water balances (Sellers, 1965, pp. 82–126). Moreover, many pre-war meteorological concepts have had to be revised. Mid-latitude depressions, for instance, are no longer regarded as accidental frills to the general circulation, but rather as an enormously important, essential mechanism of energy transfer and exchange (Hare, 1965). Similarly, fronts are no longer thought of in quite the same way as they were 10 years ago; they are now regarded as secondary consequences rather than causes of circulation patterns and not necessarily created by the juxta-position of unlike air masses. Air masses themselves have been presented in a new light allowing a greater flexibility in description which discards the rigid, source region criterion of definition and places more stress on synoptic circulation patterns (Lamb, 1965; Sutcliffe, 1966, pp. 113–18).

Briefly, current thinking is as follows (Barry and Chorley, 1968, pp. 17–134; Hare, 1965; Riehl, 1965, pp. 117–244). As a result of short-wave (solar) and long-wave (terrestrial) radiation exchanges, the earth everywhere but in polar latitudes has a net surplus of radiation energy and the atmosphere everywhere a deficit. The deficit does not vary a great deal with latitude but the surplus is at a maximum in the tropics (with a secondary minimum at the equator). Combining the latitudinal figures for the earth and atmosphere, there is a surplus of radiation energy receipt over losses between 40°N. and 35°S. and a deficit elsewhere. Thus in order to prevent constant increases or decreases of local temperature, heat energy must be transferred from the earth to the atmosphere and from low to high latitudes. The transfer is effected in the form of sensible heat, that is warm winds and thermals, as latent heat (evaporation followed elsewhere by condensation) and by ocean currents.

Many complex forces operate upon moving winds and because of these there is not a simple, single-cell circulation linking low and high latitudes (Barry, 1967, pp. 99–114; Barry and Chorley, 1968, pp. 98–134). Instead, averaged over a long period, the circulation trends are as shown in Fig. 4. The patterns at any one time will, of course, be highly complex but the simple model does enable us to understand many of the basic features of the earth's climates: the unstable Intertropical Convergence Zone, the anticyclonic subsidence limiting cloud development in tropical desert areas, the spatial

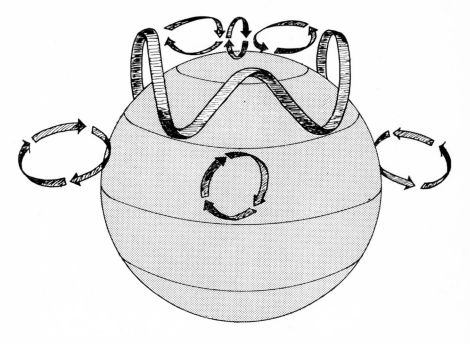

Fig. 4. A very simple model of the general circulation. In low and high latitudes, air circulates through sloping cells known respectively as the Hadley Cell and the Polar Cell, the former being much stronger and more persistent than the latter. In middle latitudes, above about 3 km, air moves through a series of superimposed waves of various wavelength. The largest stretch over 90–120 degrees of longitude; the smallest (linked at the ground to moving depressions and anticyclones) have wavelengths of 70 degrees or less.

and temporal variety of mid-latitude climates in a zone of stationary and migratory wave patterns of pressure and winds, and the ephemeral subsidence and divergent near-surface winds of the polar troposphere.

A great deal of attention has been paid in recent years to the nature of the mid-latitude tropospheric wave patterns and their relationship to surface synoptic systems (Hare, 1960; Sutcliffe, 1966, pp. 93–113). Not only present but past (Lamb and Johnson, 1959) and near-future (Sutcliffe, 1966, pp. 119–28) weather can be understood and forecast only in terms of tropospheric pressure troughs and ridges which were barely charted before 1950. Convergence and divergence of air on the western and eastern limbs respectively of troughs in the middle and upper troposphere are frequently related through sinking air on the western limb and rising air on the eastern limb to contrary movements at the ground. These movements help to explain the formation and movements of anticyclones and depressions in a way that was impossible 30 years ago. Anticyclones form where there is convergence of air in the middle and upper troposphere followed by sinking and divergence nearer the earth, whilst depressions form in areas where this pattern is reversed (Pedgley, 1962, pp. 165–9; 175–6).

Convergence also leads to a sharpening of temperature gradients as well as ascent, and these processes have been shown to be responsible for many fronts. Frontal patterns of temperatures, winds, clouds, and precipitation can thus be more realistically interpreted as the result of convergence leading to intense baroclinic zones (belts where the pressure and temperature fields intersect each other sharply) rather than the movement of one air mass relative to another. The latter, older explanation has been shown to be inadequate, one piece of evidence being the non-coincidence of frontal clouds and their associated weather phenomena with the zone of intense thermal gradient or front. Fronts are thus regarded more as dynamic phenomena than as boundaries between unlike air masses as presented in the models of the Norwegian school (Sutcliffe, 1966, pp. 117–18). Research on fronts has also shown them to be far more complex than these early models suggested. Air in the neighbourhood of cold fronts away from the depression centre is frequently subsident causing a low subsidence inversion which limits cloud development to a shallow layer near the ground. Such a front is known as a kata-cold front. Nearer the centre of the depression, the classical ana-cold front associated with strong ascent is frequently to be found. Kata-warm fronts associated with descending warm air are usually associated with only a shallow layer of low, stratocumulus cloud, whereas the ana-warm front, with rising warm air, has multi-layered cloud commonly extending through the whole troposphere (Pedgley, 1962, pp. 142–7).

Synoptic Climatology

The synoptic approach to climatology, the third section of the suggested syllabus, has revolutionized climatic description in the last few decades. Modern climatology now uses meso-scale meteorological phenomena as its units of description and analysis, analysing regional conditions in terms of the nature and periodicity of synoptic systems and their associated weather (Trewartha, 1961). Here too, analyses since 1950 of the middle and upper tropospheric circulation patterns with their series of in and out of phase stationary and migratory pressure ridges and troughs have led to fundamental changes in our understanding of many mid-latitude climates. Continentality, for instance, can no longer be satisfactorily explained in static, terrestrial–thermal terms alone. It is much more realistically thought of in a manner stressing the importance of seasonal changes in the position of pressure waves and hence tropospheric airstream trajectories in the middle and upper troposphere. This also applies to our understanding of the monsoon of south-east Asia (Barry, 1967, pp. 123–6; Barry and Chorley, 1968, pp. 226–39; Trewartha, 1958).

Local Climates

In the field of local climatology, the final section of the proposed syllabus, work has been less revolutionary but no less interesting. Indeed, atmosphere–surface interactions occupy a very important place in school geography and students commonly find this branch of the subject the most attractive and in many respects more relevant part of climatology. The results are often clear from field evidence and the approach is generally strongly geographical (Geiger, 1965; Munn, 1966). Some knowledge of the difference in climate between the standard height of measurement, 4 feet, and within a growing crop are essential if the correct interpretation is to be placed upon bio-climatological relationships and much recent research has been done within this field (*Agricultural Meteorology*, 1964). Forest climates (Food and Agricultural Organization, 1962) and urban climates, including atmospheric pollution (Chandler, 1965; Garnett, 1967; Scorer, 1968; U.S. Department of Health, 1961), have also received a good deal of attention in recent years.

CONCLUSION

In retrospect, much pre-Second World War meteorology and climatology looks remarkably naïve and unscientific. Viewed differently, one is surprised how much could be done in those days with so little knowledge of anything but near-surface conditions. The greatest revolution has been the addition, virtually for the first time in any real sense, of the third dimension. Today, even stratospheric studies, made possible by such aids as satellites and meteorological research rockets and computers, which only a short time ago were thought of as completely irrelevant to surface conditions, are now known to bear upon tropospheric processes, although at present the relationships are highly speculative (Hare, 1962).

There are few branches of climatology, except perhaps boundary-layer studies, where middle and upper tropospheric circulation patterns are irrelevant and these have been systematically charted and analysed in only the last two decades. Climatology is no longer possible without their consideration and this is the greatest single consequence of recent research in the field.

REFERENCES AND FURTHER READING

THE STATUS OF METEOROLOGY AND CLIMATOLOGY

Recent general reviews of meteorology and climatology include:

BARRY, R. G. (1967) Models in meteorology and climatology, in CHORLEY, R. J. and HAGGETT, P. (Eds.) *Models in Geography*, Methuen, London, pp. 97–144.

BARRY, R. G. and CHORLEY, R. J. (1968) *Atmosphere, Weather and Climate*, Methuen, London.

PEDGLEY, D. E. (1962) *A Course of Elementary Meteorology*, H.M.S.O., London.

RIEHL, H. (1965) *Introduction to the Atmosphere*, McGraw-Hill, New York.

SUTCLIFFE, R. C. (1966) *Weather and Climate*, Weidenfeld and Nicolson, London.

The running of a school weather station and the use of weather maps are considered in:

BAYLISS, J. M. (1962) *Running a School Weather Station* (filmstrip), Diana Wyllie, London.

HOOKEY, P. G. (1968) *Do-It-Yourself Weather Instruments*, Geographical Association, Sheffield.

TAYLOR, J. A., and YATES, R. A. (1967) *British Weather in Maps*, Macmillan, London.

RECENT ADVANCES IN METEOROLOGY

In addition to the general reviews listed above, much valuable recent research work is considered in:

MASON, B. J. (1962) *Clouds, Rain and Rainmaking*, Cambridge U.P., Cambridge.

RECENT ADVANCES IN CLIMATOLOGY

Important contributions to this rapidly growing subject include the general reviews mentioned above and also:

HARE, F. K. (1960) The westerlies, *Geographical Review*, **50**, 345–67.

HARE, F. K. (1962) The stratosphere, *Geographical Review*, **52**, 525–47.

HARE, F. K. (1965) Energy exchanges and the general circulation, *Geography*, **50**, 229–41.

HARE, F. K. (1966) *The Restless Atmosphere*, Hutchinson, London.

LAMB, H. H. (1965) Frequency of weather types, *Weather*, **20**, 9–12.

LAMB, H. H. and JOHNSON, A. I. (1965) Climatic variation and observed changes in the general circulation, parts 1 and 2, *Geografiska Annaler*, **41**, 94–134.

SELLERS, W. D. (1965) *Physical Climatology*, U. of Chicago P., Chicago.

TREWARTHA, G. T. (1958) Climate as related to the jet stream in the Orient, *Erdkunde*, **12**, 204–14.

TREWARTHA, G. T. (1961) *The Earth's Problem Climates*, McGraw-Hill, New York.

LOCAL CLIMATES

Many studies in bio-climatological relationships appear in:

Agricultural Meteorology (1964) Elsevier, Amsterdam.

The following works are a small sample of recent work on local climates:

CHANDLER, T. J. (1965) *The Climate of London*, Hutchinson, London.

Food and Agricultural Organization of the United Nations (1962) Forest influences, *Forestry and Forest Products Studies*, **15**.

GARNETT, A. (1967) Some climatological problems in urban geography with reference to air pollution, *Transactions and Papers of the Institute of British Geographers*, **42**, 21–43.

GEIGER, R. (1966) *The Climate Near the Ground*, Harvard U.P., Cambridge, Mass.

MUNN, R. E. (1966) *Descriptive Micrometeorology*, Academic P., New York.

SCORER, R. S. (1968) *Air Pollution*, Pergamon P., Oxford.

United States Department of Health (1961) Air over cities, *Technical Report*, **A62-5**.

CHAPTER 4

THE ECOSYSTEM AND
THE COMMUNITY IN BIOGEOGRAPHY

CAROLYN M. HARRISON

INTRODUCTION

Purely descriptive inventories of plant communities by early biogeographers were injected many years ago by the dynamic views of the American, F. E. Clements (1916). He introduced the concept of *plant succession*, in which plant communities are recognized as developing in complexity over a period of time until an ultimate, or *climax* vegetation is achieved. The climax vegetation is viewed as a community in equilibrium with its environment, and is hence self-perpetuating. All communities which are not in equilibrium, the *seral* communities, are developing towards a climax vegetation. Although Clements envisaged the prevailing climate as being of overriding importance in determining the character of the climax vegetation, the British ecologist A. G. Tansley (1935) stressed the importance of other environmental factors. Slope, aspect, soil, exposure, and the activities of man and animals could all prevent the climatic climax from being attained. He thus suggested a *poly-climax* theory in which climax communities were defined as those in equilibrium with all of their particular environmental conditions. In Britain, for example, although the climatic climax of lowland areas would be mixed oak forest, riverine tracts with waterlogged soils would be characterized by a climax community of alder and willow woodland, and the excessively well-drained chalklands are thought to offer a competitive advantage to beech trees.

The two concepts of plant succession and poly-climax are the cornerstones of recent developments in biogeography. Their impact rests on the importance in both of the dynamic relationships which exist among plants, animals, man, and the inanimate environment through time. It is these relationships which Tansley (1935) welded into the extremely important and fundamental concept of the *ecosystem* (Stoddart, 1965). Biogeography is now characterized by an ecological approach which seeks to demonstrate the functioning of relationships which exist between all living things and the physical environment.

36

THE STRUCTURE AND DYNAMICS OF THE ECOSYSTEM

"Ecosystem" was the term proposed by Tansley to express the "*interaction system* comprising living things and their non-living environment". Subsequent studies have emphasized that the ecosystem possesses all the characteristics of a working system, in which the sun's energy is trapped by the process of photosynthesis and forms the food source for all other life supported by the system. The transfer of energy within the system from green plant to herbivore, and herbivore to carnivore is never completely efficient, for there is a loss of heat between each feeding level. The ultimate numbers which can be supported by a system are therefore limited by the amount of energy fixed by the green plants. The inefficient transfer of energy between feeding levels imparts structure to the system.

Five feeding levels (known as trophic or energy-availing levels) are generally recognized in most ecosystems (Billings, 1964). For example in a woodland system the primary producers ($T1$ = trophic level 1) are the green plants such as trees, shrubs, and herbs. The primary consumers ($T2$) are herbivores and are represented by rabbits and numerous small mammals. The secondary consumers ($T3$) are carnivores such as the stoat, fox, and weasel. The top carnivore level ($T4$) is represented by a few birds of prey such as hawks and buzzards. Members of the highest level may feed off individuals of the lower levels, $T2$ and $T3$. Decomposer organisms, living for the most part in the soil, represent level $T5$. Soil decomposers occur in large numbers and include mites, earthworms, bacteria, and fungi. All ecosystems are structured in this way so that all green plants and animals (the *biomass*) in any system are controlled by the initial input of solar energy. In tropical areas with high values of insolation, energy may never be a limiting factor to plant growth throughout the year, although in temperate latitudes it is obviously limiting during the winter period. The vegetation of tropical areas is typically luxuriant and the tropical rain forest has a large biomass. In contrast, biomass is small in arctic and alpine tundra communities.

It is insufficient to recognize energy as the only limiting factor to the efficient functioning of ecosystems. Water and minerals are two basic inputs which must be freely available at all levels within a system if there is to be an efficient use of energy. Water enters a system as precipitation, through the subsoil by seepage from rivers, and from underground sources. It is lost from a system by evaporation from water surfaces, by transpiration from plants and animals, and by percolation through the soil. A similar cycle exists for the essential minerals in a system. The main supply needed by plants and animals for the build-up of tissues comes from the soil and, in some lower forms of plant life, from the atmosphere. When plants and animals decay, these minerals are released by the activities of decomposer organisms. In the absence of any of these basic input requirements, the efficiency of a system is impaired. In many exploited systems, such as those in which agriculture and forestry play an important rôle, minerals are removed from the system faster than they are replaced, with the ultimate result that the environment deteriorates and the system runs down. It has been suggested that for many areas of temperate agriculture, mineral deficiency is a limiting factor to the efficient functioning of the system although this is accentuated by insufficient water in more continental areas.

Figure 5 summarizes diagrammatically the dynamics of a woodland eco-system. Input factors are energy, rain water, soil minerals, and soil water. Output factors include the standing crop in the form of timber and logs, the animals dependent upon the primary producers, and water losses. Loss of heat from the system, another output factor, is not shown. The volume of material passing through such a system and the efficiency of energy exchange between the different trophic levels can be estimated and quantified (Phillipson, 1966), so that the gross primary production of a system can be measured.

Primary production is the whole dry matter production of the trees, shrubs, and ground vegetation, and is a measure of the efficiency of energy fixation of that system (Newbould, 1967). Because of their complexity, production studies of world ecosystems call for long-term observation and co-operation among a number of research workers. With this end in view, the International Biological Programme, currently in progress, was established so that comparative data could be assembled for a representative selection of the world's principal terrestrial and aquatic communities. In Britain, research workers from the Nature Conservancy and other organizations are studying the production of a mixed deciduous woodland at Meathop Wood, Grange-over-Sands, Lancashire. As a result of such studies growth rates and productivity of species can be compared in different communities of the world in both natural and man-made systems. The results will also enable a more rational development of land use and management to be made upon sound ecological principles (Nicholson, 1968). In agricultural systems, for example, some measure of economic value may be devised, and in natural and planted forests the use of L.D. Stamp's Standard Nutritional Unit (10^6 calories per year) might allow the productivity of economic species to be measured (Stamp, 1960).

Similar productivity studies can be undertaken in schools by instituting long-term projects within areally restricted ecosystems. Input and output factors for small communities can often be obtained over short periods, although in Britian seasonality of many communities demands longer term observation. Energy flow, feeding relationships, and mineral and water cycling can be established for such systems as a water tank, a fallen log, or even a brick wall. Many such suggestions for school projects are contained in

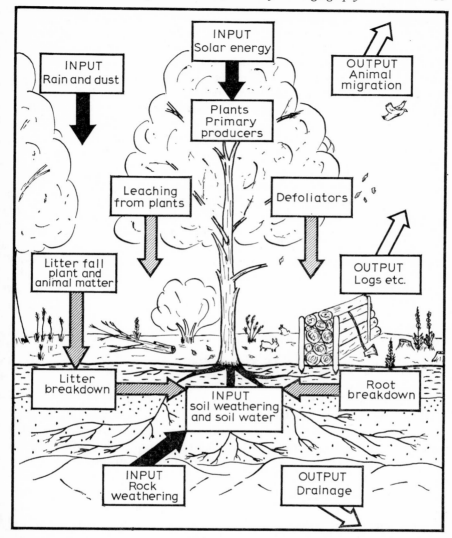

Fig. 5. Dynamics of a woodland ecosystem. (After Ovington, J. D., 1962, *Advances in Ecological Research*, **1**.)

a symposium on the teaching of ecology (Lambert, 1967). The holistic approach of ecosystem studies means that co-operation with the biological sciences is essential if complete energy budgets are to be obtained: inter-disciplinary co-operation is a particular attribute of ecosystem studies and should be encouraged at all levels of scholarship.

Studies of seral communities and the investigation of the progressive build-up of productivity with older communities are particularly relevant within the context of the ecosystem. Given an unlimited supply of essential elements and the plants and animals available to colonize an area, there is, with the passage of time, a gradual build-up of biomass and a more productive system. All seral communities are in the process of change and their productivity is restricted by the absence of one or more of the essential input requirements. In the primary colonization of bare ground, for example, there might be a shortage of minerals, and existing minerals may be cycled slowly in the absence of suitable fauna. Such a community is unstable and easily destroyed by man and animals. Erosion of young sand dunes as a result of man's activities illustrates this instability. Long-term observations over an area of recently burned ground or felled woodland, or of the colonization of bare ground such as tip heaps and sand, would indicate the progressive increase in production and the ability of older communities to resist disruption from such pressures as grazing and trampling. Exclusion of different trophic levels within the developing system would also allow the potential production of suppressed species to be assessed. For example, the exclusion of the first herbivore level, represented by rabbits, from grassland, would provide a dramatic example of their grazing effects upon the species present (Thomas, 1960).

MAN IN THE ECOSYSTEM

Man must also be considered as a component of many ecosystems, both as the instigator of new systems and as a component of different trophic levels within existing systems (Fosberg, 1963; Simmons, 1966; see also Chapter 13, below). Man in his particular rôle as an omnivore complicates existing feeding relationships between the trophic levels of a system. His failure to recognize the implications of cropping and exploitation throughout an entire system results in widespread degradation. Soil erosion resulting from overgrazing bears witness to man's indiscriminate exploitation of pasture systems in East Africa and elsewhere. The eradication of hedgerows in East Anglia must be seen not only as the loss of a valuable habitat for small mammals and birds which keep crop pests under control, but also as the loss of protective plant cover which prevents soil erosion. An awareness of the functioning of the ecosystem would have prevented the disruption of terrestrial and aquatic communities by the indiscriminate application of herbicides and pesticides, as Rachel Carson (1963) so vividly described.

Odum (1959) has shown that when man replaces existing natural systems with agricultural systems, the net production of the new system is almost always lower than that of the existing one. Monocultures such as cereal cropping and forestry are simplified systems in which the pathway of energy relations is cut short and mineral and water cycling is impeded. Overall productivity is consequently reduced. In a study of two Breckland ecosystems, Ovington (1957) demonstrated that forestry shows a two-fold increase in productivity over the most efficient cereal farming practised in that area. In this particular example, although both communities are exploited systems, it is evident that the current land use of any area is not always the most economic, or ecologically the most efficient system. Further studies of this nature are necessary for a more rational approach to land-use planning and conservation of resources.

COMMUNITY DESCRIPTION AND ANALYSIS

Such planning of land use should be made upon recommendations arising out of accurate and detailed community studies. The early descriptive studies of many ecologists relied upon a subjective assessment of dominance of individual species. Dominance was assessed on an informal scale ranging from dominant, through abundant, frequent, and occasional, to rare, with the prefix "locally" added when necessary. Such an arbitrary scale is subject to difficulties of interpretation because definitions will vary with the observer, the time of year, and the area sampled. Lack of standardization does not permit accurate comparisons to be made among several communities or between different communities through time. It is also difficult to attempt any accurate correlation of species performance with environmental factors. Thus, community units recognized on this basis cannot serve as an adequate framework within which detailed ecological studies can be undertaken. Similarly, the analogy which likened a plant community to an organism and found favour with many Americans and British workers before 1950 (Tansley, 1935), finds no place in current community studies. The organismic analogy does nothing to further our understanding of the functional relationships which exist among components of a community.

Attempts to classify vegetation into communities which are conceived as independent viable units have met with only limited success. Vegetation can never be viewed as being entirely comprised of distinct units for there are always large areas of overlap between communities described in this manner. Such areas of transition Tansley termed *ecotones* and they find only a small place in purely floristic classifications, although they may actually be quite extensive. For example, the forest-tundra ecotone in North America occurs over 200 miles of latitude. Many modern techniques of community description and analysis indicate that vegetation varies continuously in space along several environmental gradients. It is only in areas where there are marked discontinuities in these gradients, as a result, for example, of long-continued selective processes such as grazing or burning, that distinct boundaries may be found between communities. Although there has been much controversy as to whether vegetation is a continuum or not, which has tended to distract the attention of workers from the main purposes of community analysis, it is evident that with the use of objective and standardized techniques there is greater comparability between the results of several workers than was possible using the more informal descriptive methods. A convenient summary of recent methods is to be found in Kershaw (1964).

These standardized techniques have been most successfully applied to the correlation of communities with environmental factors, notably by Poore (McVean and Ratcliffe, 1962). His method is a modification of a system of community description widely employed in continental Europe and first initiated by Braun-Blanquet (Küchler, 1967). The field procedure involves a subjective selection of homogeneous stands of vegetation which are then characterized by a full inventory of all the species present. Various descriptive estimates can be used to represent the species occurrence, abundance, dominance, sociability, vitality, cover, and density.* Records of the physical site factors are also made for each stand sampled. Several lists are compiled so that the variation exhibited by the vegetation is adequately represented. In the laboratory, lists are compared according to their species composition only, and similar lists are provisionally grouped together. Mutually exclusive species may be found and used to divide initially the total number of lists into two groups. Other species may only occur together and would thus serve to indicate a third group. This process of successive approximation is based solely upon floristic criteria. It is thus axiomatic to the method that the species are the best indicators of environmental conditions and hence recurrent groups of species must be a reflection of distinct habitat controls.

The lowest community unit which Poore recognized is the *nodum* (plural, noda) and that of Braun-Blanquet is the *association*. An initial inspection of like lists will reveal that some fall readily into distinct groups while others lie transitional to one or more such groups. Transitional lists such as these are generally excluded from the Braun-Blanquet system and play no further part in the grouping. Poore, however, preferred to consider them as representing transitional communities. His work in Scotland led him to the conclusion that the most commonly occurring set of ecological conditions for that area—the noda—formed a frame of reference within which the vegetation varied continuously. Theoretically, any newly described stands from within the area, or from an area exhibiting similar ecological relationships could be inserted into this framework. The work of McVean and Ratcliffe (1962) produced such a framework for the communities of the Scottish

* For a full explanation of these terms see K. A. Kershaw (1964, pp. 9–20).

highlands and hence allows most communities in that area to be viewed in relation to the ecological gradients represented by their analysis.

The ease with which habitat criteria can be incorporated within the definition of community units is the most promising outcome of contemporary community analysis. Environmental gradients, which are revealed by floristic analysis and which can subsequently be quantified, may serve as valuable indicators for the future use of land and the development of marginal areas. The introduction of such standardized methods of description as that of Poore and Braun-Blanquet will prove an essential step towards an understanding of vegetation variation. The application to ecology of multivariate methods, which seek to isolate significantly associated groups of species, is also a particularly promising development in that the relationship between plants and the environment involves many complex and interrelated factors. Such methods as those of Williams and Lambert (Kershaw, 1956, p. 150) seek to assess simultaneously and objectively the relationships between plants and many environmental variables. They largely depend upon access to computers for the efficient processing of data. Many of the assumptions in Poore's method are, however, essentially similar to those of the more sophisticated multivariate techniques. Both approaches rely upon the representation of vegetation by abstractions from floristic lists collected in the field; both depend upon the recognition of association between groups of species to provide abstract community groups, and both enable habitat criteria to be considered as part of those community groups.

Progress in biogeographical studies in recent years has been centred on the application of the ecosystem concept, and on methods of community description and analysis. The development of standardized and quantitative procedures for interrelating vegetation with the environment and the application of these methods to future land use, planning, and conservation is of major importance to geographers. Furthermore, the concept of the ecosystem is likely to make a lasting contribution to geographical study because it provides an analytical approach to the understanding and quantification of the relationships that exist among plants, animals, man and the environment. This has previously been absent in much geographical work and an appreciation of the ecosystem in which the source of energy and other input requirements are visible and often quantifiable on any scale, will provide a profitable introduction to many other aspects of contemporary geographical research and teaching.

REFERENCES AND FURTHER READING

INTRODUCTION

The idea of plant succession is outlined in:

CLEMENTS, F. E. (1916) Plant succession, an analysis of the development of vegetation, *Publication of the Carnegie Institution of Washington*, **242**.

Poly-climax and ecosystem concepts are reviewed in:

TANSLEY, A. G. (1935) The use and abuse of vegetational concepts and terms, *Ecology*, **16,** 287–307.

STODDART, D. R. (1965) Geography and the ecological approach: the ecosystem as a geographic principle and method, *Geography*, **50,** 242–51.

THE STRUCTURE AND DYNAMICS OF THE ECOSYSTEM

The only readily available textbook which uses the ecosystem approach is:

BILLINGS, W. D. (1964) *Plants and the Ecosystem*, Macmillan, London.

A useful reference book is:

ODUM, E. P. (1959) *Fundamentals of Ecology*, 2nd edn., Saunders, Philadelphia. Abridged and reissued in 1966 as *Ecology*, Holt Reinhart, New York.

Production ecology is reviewed in:

NEWBOULD, P. J. (1967) *Methods of Investigating the Primary Production of Forests*, I.B.P. Handbook, **2,** Blackwell, Oxford.

NICHOLSON, E. M. (1968) *Handbook to the Conservation Section of the I.B.P.*, I.B.P. Handbook, **5,** Blackwell, Oxford.

PHILLIPSON, J. (1966) *Ecological Energetics*, Studies in Biology, **1,** Arnold, London.

STAMP, L. D. (1960) *Our Developing World*, Faber, London.

Suggestions for school projects are contained in:

LAMBERT, J. M. (Ed.) (1967) *The Teaching of Ecology*, British Ecological Society Symposium, **6,** Blackwell, Oxford.

A specific illustration of this theme is:

THOMAS, A. S. (1960) Changes in the vegetation since the advent of myxamatosis, *Journal of Ecology*, **48,** 287–306.

MAN IN THE ECOSYSTEM

For a general review of land use and ecology see:

SIMMONS, I. (1966) Ecology and land use, *Transactions of the Institute of British Geographers*, **38,** 59–72.

Specific illustrations of this theme are provided by:

CARSON, R. L. (1963) *Silent Spring*, Hamilton, London.

FOSBERG, F. R. (1963) *Man's Place in the Island Ecosystem*, Symposium 10th Pacific Science Congress, Honolulu, 1961, Bishop Museum P., Bishop.

OVINGTON, J. D. (1957) Dry matter production of *Pinus sylvestris* L., *Annals of Botany*, **21,** 287–314.

COMMUNITY DESCRIPTION AND ANALYSIS

The principal techniques are described in:

KERSHAW, K. A. (1964) *Quantitative and Dynamic Ecology*, Arnold, London.

Poore's method and its application to British communities are described in:

McVEAN, D. N. and RATCLIFFE, D. A. (1962) Plant Communities of the Scottish Highlands, *Nature Conservancy Monograph*, **1.**

Braun-Blanquet's method is summarized in:

KÜCHLER, A. W. (1967) *Vegetation Mapping*, Ronald P., New York.

CHAPTER 5

THE STUDY OF SOILS IN GEOGRAPHY

A. WARREN

INTRODUCTION

The dearth of acceptable general statements about soil geography makes it a difficult subject to teach in the classroom and in the field. This dearth exists in spite of the accumulation of a bewildering amount of apparently little-ordered information, which can be attributed to the fascinating variety of factors involved in soil development and to the conflicting demands of numerous associated disciplines. The geographer alone makes several distinct demands upon pedological information, and it is important in any approach to the subject first to be aware of the objectives of study, before selecting an appropriate viewpoint. For any particular viewpoint some general statements can be made, and if subsequently the requirements or the scale of the study change, another viewpoint with its own generalizations must be selected. In this chapter four different approaches within soil geography are chosen with the aim of constructing a brief general statement for each: (1) the soil and its environment, which is perhaps the best introduction to the study as a whole; (2) the closely related topic of soil profile development; (3) soil classification; and (4) soil use, which is closely linked to soil classification.

For an understanding of soil science the student needs a basic vocabulary related to such features as soil texture, structure, mineralogy, organic matter, and moisture, and a sharpened awareness of such topics as vegetation and relief. These topics are well treated in many basic textbooks on soil and can be grasped in outline without too deep a knowledge of chemistry, biology, or physics (e.g. Buckman and Brady, 1960, pp. 1–287; Comber, 1960, pp. 1–85; Duchaufour, 1965, pp. 1–146).

THE SOIL AND ITS ENVIRONMENT

After an introduction to the vocabulary, a course on soil geography might begin with a study of the relations between the soil and its environment, and this could be based on the classical general statement by Jenny (1941). Jenny saw soil properties as a function of five basic factors: climate, parent material, vegetation, relief, and time. He also allowed for other factors, of which "man" is probably the most important.

Jenny's excellent book has stimulated a large number of studies which leave his general scheme unchallenged but which have refined some of his concepts, clarified some of the relations among them, and discovered some of the mechanisms which lie behind the effects he demonstrated. An extended form of Jenny's statement is summarized diagrammatically in Fig. 6. In this diagram the factors are ranked: geology and climate are primary; geomorphological, relief, ecological, and vegetation factors are secondary; and the soil attributes themselves are tertiary. The ecological and geomorphological factors need some further explanation: they include the "chance" factors which contribute to vegetation and relief patterns, such as position within a drainage network or in a plant community or succession. "Organisms" (other than vegetation) also have a large "chance" element. The soil attributes themselves form a subsystem of their own and produce the "soil type". The time factor is not included in this diagram, but it is important to remember that the soil, the factors which create it, and the relative importance of the factors all change with time. For example, humus type may become more important with time, whereas relief may decline in importance.

There are two groups of studies which deal with soil–environment relations. The first group includes studies of simple links (Fig. 6) while the second group includes attempts at disentangling some complications (not shown on Fig. 6) which are produced within the soil system by varying intensities and periodicities among the formative processes.

Simple Relationships

The studies of these relations are of two kinds: the first comprises attempts to discover patterns of association between the factors (following Jenny) while the second includes studies aimed at discovering the dynamics of the mechanisms which produce these patterns. Some of the more important studies are briefly reviewed below; the numbers in the discussion refer to the numbered links on Fig. 6.

(1) The general relations between geology and soil can be seen by comparing maps of the Soil Survey of Great Britain with appropriate geological maps. In many of the memoirs which accompany the soil maps there are block diagrams to illustrate such relations (e.g. Mackney and Burnham, 1966, pp. 115 and 132). In addition, many recent studies have shown the subtler influences of parent material on soil properties (link (2)) such as clay mineralogy (Stephen, 1952).

(3) The link between climate and soil mineralogy is illustrated by the

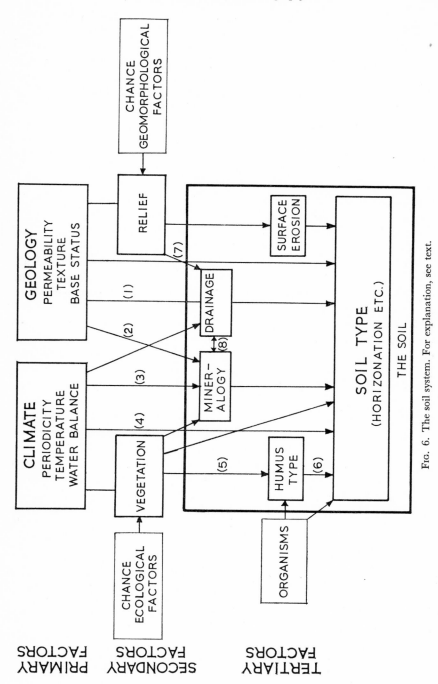

Fig. 6. The soil system. For explanation, see text.

groupings in many recent soil classifications which largely coincide with climatic zones, and by the association of these groupings with characteristic clay mineral suites. In Africa, for example, the ferralitic soils of the high rainfall areas have kaolinitic clay minerals and may even have gibbsite, whereas the ferruginous soils of dryer, more seasonal climates have kaolinites mixed with varying amounts of other minerals (D'Hoore, 1965).

Some interesting work in East Africa and Australia has revealed a relationship which involves links (3) and (4) on Fig. 6. Climate influences clay mineralogy, and thus the capacity of clay minerals to hold bases, whereas climate also influences the rate at which these bases are leached out of the soil. The amount of bases in the soil is therefore a measure of the relations between these two influences and it has been shown to be closely related to annual rainfall (e.g. Scott, 1962).

The overall effect of climate on soil type has held a fascination since Dokuchaev's time. Two examples will suffice. In the western United States the relations have been established between mean annual temperature, the amount of water available for leaching and evaporation on the one hand, and great soil groups on the other (Arkley, 1967). In Britain, studies of this link are not common but climate and soils in the Peak District have been shown to be related (Bryan, 1967).

(5 and 6). The relations among vegetation, humus type, and soil type were discussed in some detail by Duchaufour (1965, pp. 100–46 and 195–8). He showed how the carbon–nitrogen ratio of the leaf litter from various trees has an effect on humus type. In Britain the relations between the pH of a soil and the kind of trees planted in it have been studied by Ovington and Madgwick (1957).

Links (1)–(6) are the result of weathering processes. Chemical weathering was very comprehensively discussed by Keller (1962). A number of experimental studies can be found in Hallsworth and Crawford (1965).

(7) The idea that relief and soil characteristics are linked is based on the catena concept of Milne (e.g. 1944), which states that a soil's properties can be related to its slope angle and position. In parts of Scotland, for example, there is a recurring sequence of peaty podsols on upper gently sloping areas, acid brown earths on steeper valley sides, and peaty gleyed podsols in valley floors (e.g. Fitzpatrick, 1964).

(8) In many areas weathering, relief, and mineralogy may be related: for example silica may be weathered and washed out of upper slope soils and may accumulate in hollows where "high silica" montmorillonitic clays are often found (e.g. Jackson, 1965).

There are two mechanisms involved in slope–soil relationships. First, drainage is important: the processes involved in gleying, which is the result of waterlogging, were discussed by Bloomfield (1951). Second, there are slope erosion processes: in this field there is an increasing common interest among geomorphologists and pedologists which is illustrated by Beckett's discussion of the relevance of Walter Penck's ideas to soil study (Beckett, 1968).

Early studies of the effects of relatively short time periods on soil properties used British coastal dune soils (Salisbury, 1925). More recently, Crocker (1960), among others, examined soils on surfaces appearing from beneath glaciers in Alaska and on mudflows in California. Some early authorities assumed that after an initial period of fast change the soil reached an equilibrium with its environment, but more recent work has shown that in fact soils may evolve slowly over long periods of time (Cline, 1961). This is important when we realize that these long periods also include slow evolution of landforms and changes in climate. Such changes lead to complications in the relations between soil and environment which are more difficult to study.

Complex Soil–Environment Relations

The problem of finding a "good" example of a soil type has faced most people who have tried to demonstrate soils in the field; every soil profile seems to be complex. This complexity has three main sources: (1) landforms and soils are developing at rates which are not always "in phase"; (2) during the period of soil and landform development world climates have changed repeatedly (see Chapter 2, above); (3) human activity has very considerably modified the soil in most parts of the world.

(1) Geomorphological complications occur on two kinds of surfaces.

(a) Depositional surfaces. In many areas, windblown additions to the soil surface are a "normal" feature of soil development. For example, mineral particles may be blown inland from coastal dunes or beaches (Crampton, 1961). (b) Erosional surfaces. A neat generalization of the relationship between depth and age of soils in areas of parallel-retreating steep slopes has been presented by Mulcahy (1961), who showed that soils are youngest and shallowest near the base of the retreating slopes and are deeper and older on the plains farther away from them, because of the longer exposure of the more distant parts of the plains. In Sussex, Hodgeson *et al.* (1967) have shown how clay-with-flints might have evolved because of the slow stripping of the Eocene beds from the underlying chalk surface.

(2) Summaries of some of the effects of Quaternary climatic changes on soil profiles can be found in the papers of Thorp (1965), Stephens (1965), and Jackson (1965). In Australia the "K-cycle concept" (Butler, 1959), an attempt to explain different land surfaces and the soils beneath them in terms of a recurring sequence of erosional and stable phases that may be related to

climatic changes, has been found to have wide application (e.g. Walker, 1962).

British soils too have been affected by climatic changes. Indeed, few pre-glacial soils remain (Fitzpatrick, 1963), since most of these have been redistributed by glacial or periglacial activity (e.g. Kerney *et al.*, 1964). The peculiar conditions in times of intense cold seem to have produced silty loess-like layers in many soils in southern England (Avery *et al.*, 1959), and in many British soils a hard, indurated horizon can be found whose lenticular structure has led to the belief that it is a relic of soil permafrost (e.g. Cramp-ton, 1965).

(3) Man-induced modifications to soil profiles are very important in many areas (Grove, 1951; Mitchel, 1959; Sombroek, 1966). Cultivation and graz-ing can induce soil erosion, and irrigation often results in salinization. Many workers in Europe (e.g. Duchaufour, 1965, pp. 282–5) have described rem-nants of former acid-brown soils in the profiles of modern podsols, and they see this as a result of burning and cutting of former deciduous woodland, followed by the invasion of acidifying vegetation (usually heath). Such inter-ference may have a long history. Kerney *et al.* (1964), for example, have shown that there was soil erosion on the North Downs in Iron Age times.

Soil Profile Development

Mechanisms of profile development are discussed in several texts (e.g. Duchaufour, 1965, pp. 184–6; Bunting, 1965, pp. 88–100, Hallsworth and Crawford, 1965). The recent growth in the amount of information on this topic from many different parts of the world, rather than making general statements about soil profile development more difficult to construct, have shown that many processes are common to all soils, and that it is only the changing relations among these processes which differentiate one soil from another (Simonson, 1959; Cline, 1961). Duchaufour discussed the mineral-ization and humification of plant litter, and showed that while many processes are common in most environments, different end-products may result because of the varying speed of the processes, and the stages in the breakdown cycle at which decomposition has been halted. Again, Simonson (1959) pointed out the importance of clay movement down the profile in amny soils.

Because of these and other generalizations, some of the older distinctions between soil types are being blurred. In the United States, for example, the grey-brown and the red-brown podsolic soils have been described merely as younger and older expressions of the same processes, and in Alaska soils formerly termed "tundra" have been described simply as weakly-developed podsols (e.g. Tedrow *et al.*, 1958).

Using these new concepts, Simonson (1959) has suggested a generalized theory of soil development in which soil profiles are considered to be the result of four processes: additions, removals, transfers, and transformations. As Cline (1961) has indicated, such generalized models have many advantages as teaching tools.

Soil Classification

Most classifications are made for practical purposes and there seem to be few examples of classifications that are really research tools (Kubiena, 1953, p. 16, and 1958). On the other hand a classification can be an important teaching tool if its limitations are realized.

Soil classification has proved difficult, and most classifications use rather abstract concepts, because of the variety of factors which contribute to soil formation and of widespread geomorphological and climatic changes in the recent past. Many early classifications were concerned merely with "undisturbed" soils, and because these were found to be rare, the classifications only covered idealized profiles. Some recent classifications (notably the U.S. Department of Agriculture 7th Approximation, 1960) have tried to avoid inconsistencies in the older approach by introducing precise criteria for grouping. But this approach also has its disadvantages. Webster (1968a and 1968b) pointed out that the use of precise criteria can lead to soil boundaries on a map which cut across natural units, and that even in the new system there is the temptation to search for "good" examples of a soil type. Webster's most important point is that classifications should be regarded as applicable only above the soil series level, or in other words that they should deal only with idealized units and that the field surveyor should not concern himself with the place of his soils in a world scheme.

Soil classification therefore is at two levels. First, there is the field "mapping unit" or the soil series level (e.g. Mackney and Burnham, 1966), while at a completely different scale there are world classifications. Four important examples of the latter classifications are shown in Table 1. Early schemes were based principally on climatic divisions whereas the more recent schemes shown here have relied on the morphological properties of the soil itself. The U.S. Department of Agriculture 7th Approximation (1960), for example, uses diagnostic B horizons such as those rich in calcium carbonate or iron, and the "epipedon" (the upper layers of the profile); Duchaufour's classification (1965, pp. 218–25), on the other hand, stresses the humus type. Both schemes use such criteria as the degree of profile differentiation and the state of weathering.

It is apparent from Table 1 that many of the classic world soil groups recur in most classifications in spite of supposedly different approaches to the problem: the later schemes have all built on the foundations of the earlier

TABLE 1. COMPARISON OF SOME SOIL CLASSIFICATIONS

Dokuchaev–Sibirtsev	U.S.D.A. (1949)	Kubiena (1953)	U.S.D.A. 7th Approximation (1960)	Duchaufour (1965)
TUNDRA SOILS	TUNDRA SOILS (soils of the cold zone)	TERRESTRIAL RAW-MARK e.g. Rawmark; RANKER-LIKE SOILS e.g. Rutmark	ENTISOLS e.g. Cryudents; INCEPTISOLS e.g. Cryumbepts	RAW MINERAL SOILS, (A)C e.g. Polygonal arctic soils; SOILS WITH POORLY DEVELOPED PROFILES, AC e.g. Tundra soils Rankers
TAIGA SOILS (podsolized soils)	PODSOLS	PODSOL CLASS Podsols e.g. Iron, iron-humus, and humus podsols; Semi-podsols	SPODOSOLS e.g. Cryaquods Ferrods Orthods Humods	EVOLVED SOILS, ABC, WITH RAW HUMUS With and without hydromorphic mor
	PODSOLIC SOILS e.g. Grey-brown and Red-yellow Podsolic soils; (CALCIMORPHIC SOILS Brown forest soil)	BROWN EARTHS Braunerde	AFLISOLS Altaf Udalf Ustalf; SOME INCEPTISOLS Umbepts Ochepts	EVOLVED SOILS WITH A MULL HUMUS With ABC profiles e.g. *Sols bruns lessivés*; With A(B)C profiles e.g. *Sols bruns acides*
FOREST/STEPPE SOILS	FOREST/GRASSLAND SOILS Non-calcic brown soils Degraded chernozem			
STEPPE SOILS	GRASSLAND SOILS e.g. Chernozems Chestnut and chestnut-brown soils	STEPPE SOILS e.g. Chernozems; e.g. Chestnut and Chestnut-brown soils	MOLLISOLS Udolls; Altolls Ustolls; Albolls	STEPPE SOILS With ABC profiles e.g. Brunizem; With a saturated complex e.g. Chernozems; Very altered, saturated, low-humus soils e.g. Chestnut-red Red-brown soils Sub-desert soils
DESERT/STEPPE SOILS (chestnut and chestnut-brown soils)				

ZONAL SOILS

			INCEPTISOLS–Ustents	
DESERT SOILS	SOILS OF ARID REGIONS e.g. Sierozems Desert soils Red desert soils, etc.	TERRESTRIAL RAW SOILS Yerma	ARIDISOLS Orthid Argid	RAW MINERAL SOILS Desert soils
		TERRAE CALXIS BOLUS-LIKE SILICATE SOILS Braunlehm Rotlehm		SOILS RICH IN SESQUIOXIDES Red Mediterranean soils
TROPICAL AND SUB-TROPICAL SOILS	LATERITIC SOILS e.g. Reddish-brown lateritic soils laterite soils	LATOSOLS	UTISOLS Ochrult Ombrult OXISOLS VERTISOLS	Highly weathered ferruginous soils Ferralitic soils VERTISOLS
RENDZINA	CALCIMORPHIC SOILS Rendzinas	RENDZINA	MOLLISOLS Rendolls	CALCIMORPHIC SOILS e.g. Rendsinas
SECONDARY ALKALI SOILS	SALINE AND ALKALI SOILS e.g. Solonchak, Solonetz, etc.	SALT SOILS	MOLLISOLS Albolls, Aquolls ARIDISOLS Argids ENTISOLS Halaquerts	HALOMORPHIC SOILS e.g. Solonetz, Alkali, Solonchak, Solod
MOOR-MEADOW SOILS	HYDROMORPHIC SOILS Gleys	GLEY SOILS PSEUDO-GLEY SOILS	MANY SUB-ORDERS Aquert, Aquent, Aquoll, Aquod, Aqualf, Aqualt	SEVERAL SUB-GROUPS e.g. Gleys and pseudo-gleys
MOOR SOILS	Bogs	PEATS ANMOOR PEAT WARP SOILS	HISTOSOLS	HYDROMORPHIC SOILS Gleys and pseudo-gleys Peats and semi-peats
ALLUVIAL SOILS	ALLUVIAL SOILS		ENTISOLS Ustents Psamments	RAW SOILS (A)C
AEOLIAN SOILS	LITHOSOLS REGOSOLS	some YERMA TERRESTRIAL RAW SOILS	INCEPTISOLS Andepts	SOILS WITH POORLY DEVELOPED PROFILES

INTRAZONAL

AZONAL

ones. For example, the Mollisol group in the U.S. Department of Agriculture 7th Approximation is based on the Chernozem, Chestnut, and Brunizem groups of earlier classifications (U.S. Department of Agriculture, 1960, p. 32). Duchaufour (1963) has argued that the close similarity of many groupings in different systems shows that classifications are converging. Perhaps we are near to what Muir (1962) called an appreciation of the essential nature of soils. Webster (1968a), on the other hand, argued that these correspondences are possibly due to circular reasoning. If this is the case the only way out of the circle may be to use methods which have been used by ecologists in similarly complex systems (see Chapter 4, above). In Britain, for instance, Rayner (1966) has tried to classify numerically soils at the series level. These methods offer the best hope of advance, but it may be many years before a world classification can be evolved in this way. In the meantime it may be better to recommend the more "main-stream" classifications such as that of Duchaufour (1965, pp. 202–25) rather than the tongue-twisting U.S. Department of Agriculture 7th Approximation.

Accounts of world soils can be found in most textbooks (e.g. Duchaufour, 1965, pp. 226–357; Bunting, 1965, pp. 116–204) and supplementary, up-to-date accounts of some soil types were published in the *Journal of Soil Science*, **99** (1965).

Soil Use

An understanding of the ways in which a soil may be used is a link between the study of soils themselves and the rest of geography. It is in this sphere that a good system of classification can be most valuable. It is important to realize, for example, the management limitations of each world soil group: the tough sod of chernozems prevented large-scale exploitation until mechanical ploughs were evolved; some similarly intractible but very fertile tropical black soils are only now being developed (Davies, 1964); infertility of some African red soils meant that shifting cultivation was often the best response of the local peasant (Allan, 1965); ease of working and fertility of many central European loess soils led many early cultivators to settle in loess areas (Vink, 1963); important restraints should be observed in the cultivation of some Amazon soils (Sombroek, 1966); and difficulties arise when soils are over-irrigated in the tropics (Mitchel, 1959).

Soil survey maps now cover many parts of the world. In the United States and in most European countries the purpose of the soil survey is to produce a map of "genetic" soil types, from which it is hoped most kinds of practical information can be deduced. The steps between the mapping and its interpretation are not always simple and a body of interpretive literature is appearing. Vink (1963), for example, was concerned to formalize this process, and Bartelli *et al.* (1966) approached the soil map from the point of view of the urban and rural planner. In the United States there has always been an interest in producing land capability maps from the soil survey sheets (e.g. Klingbiel and Montgomery, 1961), and in Britain these techniques have now been adapted to local conditions. Mackney and Burnham (1966), for example, have prepared a map of land capability for the Church Stretton area. They distinguished seven classes ranging from land with no limitations to plant growth (Class 1) to land with severe limitations which can only be used for grasing and forestry (Class 7). These interpretations have been very much helped by some recent work into the relations among various soil characteristics. Mackney and Burnham, for example, used the formulae of Salter *et al.* (1966) to discover the moisture characteristics of soils from their textural qualities.

REFERENCES AND FURTHER READING

INTRODUCTION

The following general textbooks give good introductions to general soil concepts and vocabulary:

BUCKMAN, H. O. and BRADY, N. C. (1960) *The Nature and Properties of Soils*, 6th edn., Macmillan, New York.
COMBER, N. M. (1960) *An Introduction to the Scientific Study of the Soil*, 4th edn., Arnold, London.
DUCHAUFOUR, P. (1965) *Précis de Pédologie*, 2nd edn., Masson, Paris.
JACKS, G. V. (1954) *Soil*, Nelson, London.
MILLAR, C. E., TURK, L. M. and FOTH, H. D. (1965) *Fundamentals of Soil Science*, 4th edn., Wiley, New York.

THE SOIL AND ITS ENVIRONMENT

The following textbooks give general treatments of this topic:

BUNTING, B. T. (1965) *The Geography of Soil*, Hutchinson, London.
JENNY, H. (1941) *Factors of Soil Formation*, McGraw-Hill, New York.

Works referred to in connection with simple soil-environment relations are:

ARKLEY, R. J. (1967) Climates of some great world soil groups of the western United States *Soil Science*, **103**, 389–400.
BECKETT, P. H. T. (1968) Soil formation and slope development, I: a new look at Walter Penck's *Aufbereitung* concept, *Annals of Geomorphology*, **12**, 1–24.
BLOOMFIELD, C. (1951) Experiments in the mechanics of gley formation, *Journal of Soil Science*, **2**, 196–211.
BRYAN, R. (1967) Climosequences of soil development in the Peak District of Derbyshire, *East Midlands Geographer*, **4**, 251–61.
CLINE, M. G. (1961) The changing model of the soil, *Soil Science Society of America Proceedings*, **25**, 442–6.
CROCKER, R. L. (1960) The Plant factor in soil formation, *Proceedings of the Ninth Pacific Science Congress, Bangkok, Thailand*, **18**, 84–90.
D'HOORE, J. L. (1965) *Soil Map of Africa, Explanatory Monograph*, Joint Project No. 11, Commission for Technical Co-operation in Africa, Lagos.

DREW, J. V. (Ed.), (1967) *Selected Papers in Soil Formation and Classification,* Soil Science Society of America, Special Publication, **1,** Madison, Wisconsin. This collection includes the papers of Simonson, Stephen, Tedrow and others listed here, and many other useful papers.

FITZPATRICK, E. A. (1964) The soils of Scotland, in BURNETT, E. H. (Ed.), *The Vegetation of Scotland,* Oliver & Boyd, Edinburgh.

HALLSWORTH, E. G. and CRAWFORD, D. V. (1965) *Experimental Pedology,* Butterworth, London.

JACKSON, M. L. (1965) Clay transformation and soil genesis during the Quaternary, *Soil Science,* **99,** 15–22.

KELLER, W. D., (1962) *The Principles of Chemical Weathering,* Lucas, Columbia.

MACKNEY, D. and BURNHAM, C. P. (1966) *The Soils of the Church Stretton District of Shropshire,* Memoirs of the Soil Survey of Great Britain, H.M.S.O., London.

MILNE, G. (1944) Soils in relation to native population in Usambara, *Geography,* **29,** 107–13.

OVINGTON, J. D. and MADGWICK, H. A. (1957) Afforestation and soil reaction, *Journal of Soil Science,* **8,** 141–9.

SALISBURY, E. J. (1925) Notes on the edaphic succession in some dune soils with special reference to the time factor, *Journal of Ecology,* **13,** 322–8.

SCOTT, R. M. (1962) Exchangeable bases of mature well drained soils in relation to rainfall in East Africa, *Journal of Soil Science,* **13,** 1–9.

STEPHEN, I. (1952) A study of rock weathering with reference to the soils of the Malvern Hills, *Journal of Soil Science,* **3,** 20–33.

Complex soil–environment relations are explored in:

AVERY, B. W., STEPHEN, I., BROWN, G. and YAALON, D. H. (1959) The origin and development of brown-earths on clay-with-flints and coombe deposits, *Journal of Soil Science,* **10,** 177–95.

BUTLER, B. E. (1959) Periodic phenomena in landscapes as a basis for soil studies, *C.S.I.R.O., Australian Soil Publication,* **14.**

CLAYDEN, B. and MANLEY, D. J. R. (1964) The soils of the Dartmoor granite, in SIMMONS, I. (Ed.), *Dartmoor Essays,* the Devonshire Association, Torquay, pp. 117–40.

CRAMPTON, C. B. (1961) An interpretation of the micromineralogy of some Glamorgan Soils, *Journal of Soil Science,* **12,** 158–71.

CRAMPTON, C. B. (1965) An indurated horizon in soils of South Wales, *Journal of Soil Science,* **16,** 210–29.

FITZPATRICK, E. A. (1963) Deeply weathered rock in Scotland, its occurrence, age and contribution to soils, *Journal of Soil Science,* **14,** 13–42.

GROVE, A. T. (1951) Soil erosion and population problems in southeast Nigeria, *Geographical Journal,* **117,** 291–306.

HODGESON, J. M., CATT, J. A. and WEIR, A. H. (1967) The origin and development of the clay-with-flints and associated soil horizons on the South Downs, *Journal of Soil Science,* **18,** 85–102.

KERNEY, M. P., BROWN, E. H. and CHANDLER, T. J. (1964) The Late-Glacial and Post-Glacial history of the chalk escarpment near Brook, Kent, *Philosophical Transactions of the Royal Society of London,* series B, **248,** 135–204.

MITCHEL, C. W. (1959) Investigations into the soils and agriculture, Lower Diyala area, eastern Iraq, *Geographical Journal,* **125,** 390–7.

MULCAHY, M. J. (1961) Soil distribution in relation to landscape development, *Annals of Geomorphology,* **5,** 211–25.

SOMBROEK, W. G. (1966) Amazon Soils, Center for Agricultural Publications and Documentation, Wageningen, Holland.

STEPHENS, C. G. (1965) Climate as a factor of soil formation through the Quaternary, *Soil Science,* **99,** 9–14.

THORP, J. (1965) The nature of the pedological record of the Quaternary, *Soil Science,* **99,** 1–8.

WALKER, P. H. (1962) Soil layers on hillslopes: a study at Nowra, New South Wales, *Journal of Soil Science,* **13,** 167–77.

SOIL PROFILE DEVELOPMENT

SIMONSON, R. W. (1959) Outline of a generalized theory of soil genesis, *Soil Science Society of America Proceedings*, **23**, 152–6.

TEDROW, J. C. F., DREW, J. V., HILL, D. E. and DOUGLAS, L. A. (1958) Major genetic soils of the Arctic slope of Alaska, *Journal of Soil Science*, **9**, 33–45.

YAALON, D. H. and YARON, B. (1966) Framework for man-made soil changes—an outline of metapedogenesis, *Soil Science*, **102**, 272–83.

SOIL CLASSIFICATION

DUCHAUFOUR, P. H. (1963) Soil classification: a comparison of the American and French systems, *Journal of Soil Science*, **14**, 149–55.

KUBIENA, W. L. (1953) *The Soils of Europe*, Murby, London.

KUBIENA, W. L. (1958) Classification of soils, *Journal of Soil Science*, **9**, 9–19.

MUIR, J. W. (1962) The general principles of classification with reference to soils, *Journal of Soil Science*, **13**, 22–30.

RAYNER, J. H. (1966) Classification of soils by numerical methods, *Journal of Soil Science*, **17**, 79–92.

United States Department of Agriculture, Soil Survey Staff, Soil Conservation Service, 1960, *Soil Classification, a comprehensive system, 7th Approximation*, U.S. Government Printing Office, Washington, D.C.

WEBSTER, R. (1968a) Fundamental objections to the 7th Approximation, *Journal of Soil Science*, **19**, 354–66.

WEBSTER, R. (1968b) Soil classification in the United States: a short review of the Seventh Approximation, *Geographical Journal*, **134**, 394–6.

SOIL USE

ALLAN, W. (1965) *The African Husbandman*, Oliver & Boyd, Edinburgh.

BARTELLI, L. J., KLINGBIEL, A. A., BAIRD, J. V. and HEDDLESON, M. R. (Eds.), (1966) *Soil Surveys and Land Use Planning*, Soil Science Society of America and American Society of Agronomy, Wisconsin.

DAVIES, H. R. J. (1964) An agricultural evolution in the African tropics: the development of mechanised agriculture on the clay plains of the Republic of Sudan, *Tidjschift voor Economische en Sociale Geografie* **55**, 101–8.

KLINGBIEL, A. A. and MONTGOMERY, P. H. (1961) Land capability classification, U.S. Department of Agriculture, Soil Conservation Service, *Agriculture Handbook*, **210**.

SALTER, P. J., BERRY, G. and WILLIAMS, J. G. (1966) The Influence of texture on the moisture characteristics of soils, *Journal of Soil Science*, **17**, 93–8.

VINK, A. P. A. (1963) Soil Survey in relation to agricultural productivity, *Journal of Soil Science*, **14**, 388–401.

CHAPTER 6

PROGRESS IN HYDROLOGY

G. P. JONES

THE SCOPE OF HYDROLOGY

This chapter attempts to summarize the progress being made in Britain in hydrology against the background of world-wide developments.

Hydrology has been defined in numerous ways, each of which amplifies or qualifies its literal meaning—"the study of water". The widest interpretation includes the fields of meteorology and oceanography, whereas a more restricted view limits the subject to the study of the water of the earth's land areas. The International Association of Scientific Hydrology recognizes four distinct branches: (1) surface water, (2) ground water, (3) snow and ice, and (4) limnology (the study of lakes); and of these, the first two have received the greatest attention. As a branch of earth science, hydrology encompasses the history of the hydrological cycle. The complex sequence of events in this system of circulation is illustrated in Fig. 7.

A more conventional approach is to regard hydrology as that discipline which deals with the potable water resources of the earth, their occurrence, circulation, and distribution, their physical and chemical properties, and their interactions with the physical and biological environment, including their responses to human activity. Such a lengthy description gives some indication of the wide diversity, not to say clash, of interest in the field, and identifies the unifying theme of water resources. An understanding of the basic concepts of hydrology is implicit in any study of water resources and their proposed management for the optimum good of mankind.

HISTORICAL DEVELOPMENT

Hydrology is a young science, brashly extending its boundaries into the scientific disciplines from which it takes its strength. It was created by drawing upon the general aspects of those disciplines that traditionally served the water-supply industry. Use brought refinement and a growth of specialized sectors within the major disciplines, and to these have been added the powerful analytical tools provided by the mathematical sciences.

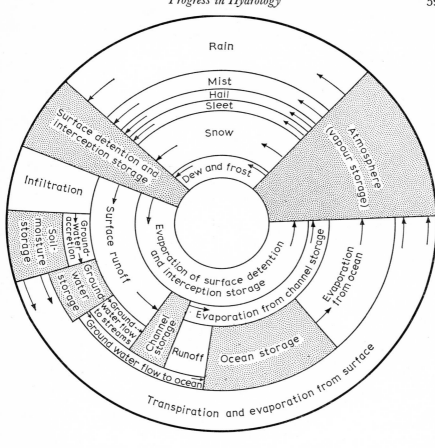

Fig. 7. The hydrological cycle. (After Wisler, C. O. and Brater, E. F., 1959, *Hydrology*, 2nd edn., Wiley, New York.)

Being such a new science it has not yet received universal recognition, either as a scientific discipline in its own right or indeed as a profession. For too long, work in hydrology has been undertaken by engineers, geologists, geomorphologists, geographers, meteorologists, chemists, and others, each looking at the subject from the viewpoint of his own individual background, and dividing the subject into separate fields of study. Clearly there has been development, but by and large it was not until the 1960's that one could detect a change of approach, with the gradual replacement of parallel development in different subjects by integrated development.

Table 2 illustrates stages in the evolution of hydrology and the increasing tempo of such evolution. Up to 1930 the construction of great hydraulic structures as a means of water management was largely accomplished by use of empirical concepts formulated in an age of qualitative hydrology. The period from 1930 to 1950, however, saw the introduction of scientific hydrology, and greater advances have been made in the last four decades than during the whole of the subject's previous history.

TABLE 2. HISTORICAL DEVELOPMENT IN HYDROLOGY (modified from Chow, 1964)

Period	Age
A Speculation (Ancient–1400) B Observation (1400–1600) C Measurement (1600–1700) D Experimentation (1700–1800) E Modernization (1800–1900) F Empiricism (1900–30)	Heroic (Ancient–1930)
G Rationalization (1930–50)	Classical (1930–50)
H Theorization (1950–60) I Application (1960 to date)	Modern (1950 to date)

The keystone of scientific hydrology is its quantitative nature and it follows that the collection, analysis, interpretation, and presentation of quantitative data is the primary concern of the hydrologist.

The quickening of interest in scientific hydrology in Britain was stimulated by the Water Resources Act 1963 which came into operation in April 1965 shortly after the initiation of the International Hydrological Decade (I.H.D.) in January of the same year. The major function of the Water Resources Act 1963 is to promote the conservation and proper use of surface-water and ground-water resources. This is done through twenty-nine authorities which are advised by the central Water Resources Board. The latter is also responsible for the long-term regional planning of water resources and the publication of systematically-collected hydrological data. One of the duties of each river authority is to carry out a survey of the water resources of its area, and such an assessment is to be followed by periodic revisions. The hydrologic variables which are measured in such surveys include precipitation and streamflow, the estimation of evaporation and the observation of ground-water level fluctuations.

In many ways the rôle of the International Hydrological Decade is to do on a world-wide scale what the river authorities are required to undertake in England and Wales. The establishment of the I.H.D. was based on the recognition that, first, many of the problems affecting human welfare in less-advanced countries of the world involve water development and management; second, that the only basis for effective water resources management is hydrological information; and third, that the hydrological cycle is not restricted by national boundaries, and so any useful study or assessment of the components of the cycle requires international co-operation.

The general programme of the I.H.D. embraces all aspects of hydrology; but the main aim is to undertake the standardized collection, analysis, and interpretation of basic hydrological data, so that "water balances" can be determined for areas ranging in size from small basins to whole continents. Scientific research into a series of problems of great theoretical and practical significance is proceeding concurrently with such inventory studies. Since there is a shortage of trained personnel to undertake the work, the education and training of hydrologists has a very important place in the programme. The I.H.D. is an international exercise in scientific co-operation, and its success will be judged as much by the continuance of such co-operation after 1974 as by the enormous mass of scientific information that will have been accumulated during the decade itself.

ADVANCES IN INSTRUMENTATION AND DATA HANDLING

One of the essential prerequisites in scientific hydrology is the development of equipment to enable the reliable collection of precise basic data under a wide range of climatic conditions. The recent increase in demand for automatically measured and recorded hydrological data and its subsequent processing has led to the development of automatic electronic instrument systems for use in the field. There are obvious advantages to be gained from a degree of standardization of components comprising the various systems, and such agreement will be beneficial to all users.

Automatic punched-tape recorders are replacing or supplementing continuous autographic pen recorders for the measurement of river-stage and ground-water levels. In complementary fashion, automatic weather stations can provide a magnetic-tape record of rainfall and those meteorological factors—solar radiation, net radiation, air temperature, temperature depression of wet bulb, wind run at a height of 2 metres, and wind direction—necessary for the determination of evaporation and, where the albedo of the surface is known, for the estimation of transpiration from a given vegetated surface.

As a check on the measurement of evaporation from automatic stations the data can be correlated with results from a variety of evaporation pans or tanks. The estimation of transpiration by formulae methods can be no more than an approximation, but with the first large hydraulic weighing lysimeter in Britain currently undergoing field trials at the Institute of Hydrology, Wallingford, it may soon be possible to compare the accuracy of the alternative methods.

Computers provide the only feasible means of handling, storing, and processing the vast amount of data that result from automated sampling. Initial problems in the provision of facilities for the translation of paper or magnetic tape into a form suitable for computer input have now been overcome and programmes for many standard operations are now in use.

SURFACE-WATER HYDROLOGY

Surface-water hydrology is largely concerned with the collection, processing, and use of river-discharge data as part of water-management schemes. Telemetry systems for receiving and recording data from remote, hydrologically significant stations are nowadays in general use for the evaluation of catchment conditions and the subsequent determination of appropriate operating procedures.

The future pattern of events seems to be that rivers will supply an increasing proportion of the growing water demand; additionally, the rivers themselves will be used instead of pipelines as a means of transferring water from one area to another. The essential control for such schemes requires a radical change in the way rivers are used. Traditionally, reservoirs have been built to provide water storage from which water undertakings abstract directly. The view currently in favour, however, is that the reservoirs should be river-regulating devices, holding or releasing water so as to maintain a steady discharge downstream and thus avoid the opposing excesses of flood flows and dry-weather flows. Prototypes of these reservoirs in Britain are the Clywedog Reservoir on the River Severn headwaters and the Llyn Celyn–Bala Lake project on the River Dee.

It is now generally accepted that much greater benefit results from schemes which serve several purposes. For instance, a multi-purpose reservoir may provide, with only occasional conflict of interests, any combination of the following services: water supply, irrigation, flood control, hydro-electric power, fishing, recreation, and navigation.

The design criteria for future schemes will require an abundance of information and among the works being undertaken in order to provide this requisite are fundamental mathematical studies for the representation of drainage basin response to precipitation. Research is also being carried out on the generation of "synthetic" river-flow data and the development of flow-regulation strategies, as well as on the estimation of the reliability of yield from reservoirs fed by pumping from rivers.

Hydrological data are the only source of information upon which scientific water-resources investigations can be based and it is fortunate that the large amount of required data is highly amenable to statistical analysis. For instance, one of the most important techniques is probability analysis, in which a past record of hydrologic events is used to express future probabilities of occurrence, e.g. the prediction of floods or droughts.

GROUND-WATER HYDROLOGY

It is becoming increasingly obvious from a water-resources viewpoint that underground and surface waters must be used as one supply; but for integrated use to be practised successfully, one needs to know as much about the first source as the second. Unfortunately the task of determining existing or potential surface-water storage is relatively easy by comparison with the assessment of those resources that are wholly hidden underground.

Because the techniques for ground-water study have remained qualitative longer than in other branches of hydrology, greater progress in quantification has been necessary. Development of new instruments has helped hasten this process and great strides have been made in the determination of important aquifer properties such as permeability and storage from both empirical formulae and laboratory measurement. Nevertheless, difficulties remain, and basic among these is the fact that natural aquifer materials are very different from the idealized homogeneous media commonly assumed in the derivation of the formulae.

One tool that has proved to be valuable in the analysis of aquifers depends on the correspondence between the flows of electricity and water. Use of electrical-analogue models is becoming commonplace for the representation of aquifers and their response to variable pumping régimes. Unfortunately, it is not possible to distinguish in electrical terms between liquids of differing density such as fresh and salt water and, for those studies involving saline-infiltration problems, recourse has to be made to viscous-flow hydraulic models.

The potential for augmenting ground-water resources by large-scale artificial recharge of aquifers has been clearly shown in the United States, Israel, and many countries in Europe, but it is only now being investigated in Britain by experimental work recently initiated in the Trent and the Thames river basins. Some time must necessarily elapse before the results of these experiments can be fully evaluated; artificial recharge on a routine basis seems to be many years away in Britain.

More imminent, however, is the introduction of schemes for phased abstraction of ground water, pilot versions of which are already being studied in the Thames and Great Ouse river basins. In the case of the Lambourne Valley (in the Thames basin), for instance, the plan is to draw upon underground water during the drier summer months and cease pumping during the wetter winter months when the aquifer would normally be replenished by infiltration. The result should be to supplement river discharge when the flow of the Thames is insufficient to meet the demand made upon it in the lower reaches, and at the same time use the river instead of a conventional and expensive pipeline for the large-scale transfer of water. The plan will have the advantage of simultaneously providing an increased supply of water for Metropolitan London at relatively low cost (a saving in both money and good agricultural land which would otherwise be needed to construct a conventional surface storage reservoir) and a more than adequate river flow for the dilution of sewage effluent.

Another interesting development is the increasing use of radio-chemical techniques, both for the tracing of ground-water movement and for the dating of ground water. To some extent rapid progress in the former is impossible given an understandable public objection to the addition of radioactive materials to those aquifers that sustain public water supplies. On the other hand, if the tracer is an inherent natural part of the ground water, as in the case of tritium and carbon-14, then these techniques provide useful tools for determining the relative age of ground waters.

CATCHMENT STUDIES

A logical way of integrating individual aspects of the hydrological cycle is to study the occurrence, distribution, and change of all components within the framework of a naturally defined areal unit—the drainage basin or catchment. The boundaries of river authorities in Britain are based on large river basins or groups of smaller ones; and an inventory of water resources within these units will be part of the hydrological surveys required by the Water Resources Act 1963. Their size and degree of modification by man make most of these basins unsuitable for a complete scientific study of the interrelationships of the hydrological cycle; hence a number of relatively small experimental catchments have been established and are taken to be representative of the larger basins.

Clearly, the detailed work in small experimental catchments requires a closer network of meteorological and hydrological stations than is envisaged for the larger basins. In the broadest sense the aim of the catchment study is to establish a water balance for the area so that the determination of individual components of the cycle may be checked for accuracy and consistency. Basically, the study will assess precipitation, streamflow, evaporation, infiltration, and the changes in storage, so as to quantify each item shown in Fig. 8 and to obtain an annual balance between input (precipitation) and the various outputs (runoff, etc.). The diagram is simplified but the complexity of the problem is clear when one superimposes onto the natural cycle the artificial "shorts cuts" that have been introduced by man in the form of

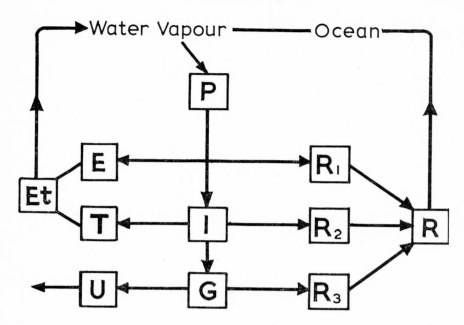

P - Precipitation E - Direct evaporation
T - Transpiration Et - Evapotranspiration
I - Infiltration G - Ground water
U - Underflow R - Streamflow
R_1 - Surface runoff R_2 - Interflow
R_3 - Ground-water component of riverflow

Fig. 8. Water movement and relations between components in the hydrological cycle.

reservoirs, river regulation, pumping from wells, land drainage, urbanization, irrigation, and vegetational changes.

The main aim behind some catchment studies is prediction of streamflow, while in others it is to determine the relationship between altitude and meteorological factors or to assess the hydrological differences between contrasting geological, topographic, and land-use conditions. For instance, the work of the Institute of Hydrology in the Plynlymon area (Central Wales) is designed to provide a long-term comparative assessment of the hydrological consequences of afforesting one catchment while leaving a comparable adjacent catchment for sheep grazing.

Complementary to the field study of catchments is work at Imperial College London, on a sophisticated laboratory model of a catchment which will provide a full appreciation of the hydrological processes involved in the transformation of rainfall into streamflow. The results of the laboratory studies will be compared and correlated with those from field studies, and it is hoped that the main practical benefit will be in the forecasting of flood conditions and the routing of flood water.

FUTURE TRENDS

Britain has a mean annual rainfall in excess of 36 in., and if one allows for more than half to be lost by evapotranspiration there still remains the equivalent of some 20,000 million gal. per day, of which we use daily no more than 4000 million gal. In this sense at least there is no shortage of water in Britain. However, when one recognizes that the greatest availability is in upland areas of the north and west and the largest need is in the southeast in summer, then the real problem is seen to be one of distribution in time and space. Some of the schemes outlined above will help ameliorate the problems, but to satisfy long-term demand one must look further ahead and take advantage of the recent fundamental change in thinking that has invigorated the field of water research. For example, artificial recharge of aquifers and such practices as irrigation and regulation of reservoirs are forms of artificial modification of the hydrological cycle which may be of immense value. Again, since precipitation is the necessary prerequisite of any scheme for water-resources development, one must seriously consider the possibility that "artificial" rain can produce the desired results. Desalinization of brackish or sea water is an available process—at a price; future application in Britain depends on the relations between the cost of using nuclear reactors for large desalinization plants and the cost of using fresh water from alternative sources. A more viable "new" idea is that of using barrages, and feasibility studies have already been commissioned for such barrages across Solway Firth, Morecambe Bay, and the Wash. In view of the enormous costs associated with these large schemes, one cannot accept any element of risk about their success. This means that the hydrological investigations must be precise and it naturally follows that they must be based on an abundance of hydrologic data of the right kind, collected in the approved manner, and analysed by trained hydrologists.

In this fashion hydrology has come of an age and, while in Britain it still has no formal place in education, there can be no doubt that the urgency regarding the development of our water resources has served to draw together the diversity of interests and to enable the presentation of a unified front which augers well for the future.

EDUCATION AND TRAINING

The need for trained personnel is felt at all levels in hydrology, from the specialists working in particular fields to the technicians responsible for collecting basic data. The continuing advance of the science is fundamentally dependent on an expansion in the output of "hydrologists", and in this context it is worth noting that the Report of the Natural Environment Research Council for 1967–8 advocated "the expansion of undergraduate teaching and postgraduate research in hydrology at universities".

At the undergraduate level there has certainly been some expansion in the teaching of hydrology, particularly in departments of geography and civil engineering, and to a lesser extent in departments of geology and earth sciences. But this expansion should be put in its true perspective since it is either a relatively small part of the required coursework or it is an optional subject taken during part of the student's final year.

The most common means of acquiring a training in hydrology is at the post-graduate level through the medium of one-year advanced courses leading to a diploma or an M.Sc. degree, and such courses as exist in Britain are tabulated below:

Name of course	University	Department
Engineering hydrology	Imperial College London	Civil Engineering
Hydrogeology	University College London	Geology
Hydrology	Newcastle-upon-Tyne	Civil Engineering
Water-resources technology	Birmingham	Civil Engineering
Water resources	Newcastle-upon-Tyne	Civil Engineering

From what has been written above it might be thought that hydrology is too diverse and too advanced to be successfully taught in schools. There is indeed more than an element of truth in such a view, if only because various modifications by man grossly complicate what is in any event only rarely a simple hydrological circulation. Nevertheless, there are some aspects which afford fertile ground for group study in schools. School weather stations have been mentioned in Chapter 3; the measurements made might be extended to include estimation of evapotranspiration and measurement of soil moisture because these are quantities in the hydrological cycle capable of satisfactory measurement in schools (World Meteorological Organization, 1966).

A most useful way in which school projects can help hydrological investigations is in the collection of basic ground-water data. For instance, the periodic measurement on a regional basis of water levels in wells and boreholes would provide much valuable information. The simple electrical instruments required for the deeper and less accessible wells may be constructed at any school with normal workshop facilities (Gray, 1961). Weekly or monthly measurements will enable the determination of seasonal fluctuations by identification of maximum and minimum water levels. From such point measurements, interpolation of the ground-water contours enables the three dimensional form of the water table to be derived and this in turn allows the boundaries of the ground-water drainage areas to be defined and the pattern of ground-water movement to be determined.

Once natural catchment areas have been delineated, the relatively simple circulation of water in the rare "undeveloped areas" can be contrasted with the complexity of water circulation in urbanized areas. With the assistance of advice and data from river authorities, a local water-resources study project can be initiated to identify the circulation of this most important, but least acknowledged, natural resource from the time it falls as precipitation, through the separate but sometimes repeated phases of the hydrological cycle, until it finally leaves the area of study in river discharge or by other means.

Hydrology, like geography itself, clearly links a number of sciences. And hydrology is the only subject which can successfully interrelate the variety of components making up the hydrological cycle. A first-hand appreciation of the problems and services provided by water supply, sewage, land drainage, and flood-protection schemes will emphasize the close dependence of society on the water cycle and the extent to which the circulation is modified by man.

REFERENCES AND FURTHER READING

THE SCOPE OF HYDROLOGY

Many books consider selected aspects of hydrology but few provide a balanced assessment of the subject as a whole, and they are, with one exception, American in origin and outlook:

Chow, V. T. (1964) *Handbook of Applied Hydrology*, McGraw-Hill, New York.
Meinzer, O. E. (Ed.) (1942) *Hydrology*, Dover, New York.
Ward, R. C. (1967) *Principles of Hydrology*, McGraw-Hill, London.
Wisler, C. O. and Brater, E. F. (1959) *Hydrology*, 2nd edn., Wiley, New York.

The aims of the International Hydrological Decade are considered in:

Batisse, M. (1965) Launching the Hydrological Decade, *New Scientist*, **425,** 38–40.
Nace, R. L. (1964) The International Hydrological Decade, *American Geophysical Union Transactions*, **45,** 413–21.

The scope and development of hydrological work in Britain is summarized in the annual reports of the *Natural Environment Research Council* and the *Water Resources Board*, published by H.M.S.O., London.

Among the many periodicals that deal wholly or partly with subjects of hydrological interest are:

Bulletin of the International Association of Scientific Hydrology.
Journal of Hydrology.
Water Resources Research.

FUTURE TRENDS

Britain's future water policy is considered in:

ANON., (1966) Finding enough water for the drift to the south-east, *Engineering*, February, pp. 338–42.

EDUCATION AND TRAINING

Techniques for measuring ground-water levels are described in:

GRAY, D. A. (1961) The measurement of ground-water levels, *Water and Water Engineering*, October, pp. 431–7.
World Meteorological Organization (1966) Measurement and estimation of evaporation and evapotranspiration, *Technical Note*, **83**.

A valuable review of certain aspects of hydrology is:

MORE, R. J. (1967) Hydrological models and geography, in CHORLEY, R. J. and HAGGETT ,P (Eds.) *Models in Geography*, Methuen, London, pp. 145–85.

CHAPTER 7

THE TEACHING OF FIELDWORK AND THE INTEGRATION OF PHYSICAL GEOGRAPHY

E. H. BROWN

INTRODUCTION

Geography is a science which has deep roots in man's intense curiosity concerning the nature of remote places. It capitalizes upon what the late S. W. Wooldridge referred to as "other-where-itis", an urgent desire to see and know what the rest of the world looks like beyond the geographical confines of our daily lives. It reflects two very deep-seated, but diametrically opposed, human beliefs: on the one hand that the grass on the other side of the fence is greener, and on the other that there is no place like home. It springs from the same sources as the desire to reach the poles, to climb Everest, and to set foot on the moon. At a more mundane level, holidays abroad and country walks are also manifestations of the same need.

There are two principal ways in which this curiosity about places may be satisfied. One line of approach is through accounts of other men's travels, their books, maps, photographs, and films—in short, the devices of the armchair traveller. These may be the secondary sources of the academic geographer. Second, we make our own primary explorations. To geographers this is "fieldwork" and it is central to the first-hand acquisition of much geographical data. There are many types of fieldwork. The most familiar adopts a "look and see" approach and is most often conducted in the form of a professional guided tour. Alternatively an individual may arm himself with a written guide of the kind produced for the London area by Wooldridge and Hutchings (1957). A second type of field work involves more active participation by the student, who is taught how to look at geography in the field from a professional point of view, to analyse the surface of the earth, to measure it, and to record the results. A great deal of this type of fieldwork is carried out on our school and university field classes and, in Britain, at the field study centres. It has the advantage of introducing a student to problems of identification, delimitation and measurement of phenomena, and involves a large number of short project studies such as the measurement of slope profiles and the mapping of agricultural land use. There is a third type of fieldwork which is primarily carried out by research workers who go into the field in order to acquire primary data for the furtherance of geographical knowledge. Whilst all teachers are familiar with the "look and see" type of fieldwork and the need for it, whether it be through the organization of a field class in their home country or abroad, and a large number (but not all) are aware of the need for project work, few are prepared to take the time and devote the energy to research ends, although they are well placed to do so. But before developing this theme a few comments on the conduct of "look and see" fieldwork and project work are appropriate.

"LOOK AND SEE" FIELDWORK

"Look and see" fieldwork gains immensely if it is linked with map reading, as the pupil is taught from the start to link what he sees on the ground with the way in which it is represented on the map. For this purpose maps on scales of between 1 : 20,000 and 1 : 75,000 are ideal. The teacher is well advised to concentrate in the field on what can be seen directly by the pupil and to reduce to a minimum the imparting of information which is not self-evident to the pupils' own eyes. There is little purpose in gathering a class together in the field simply to lecture to them on matters which are not to be seen in front of them, for they may quickly lose interest in what is being taught and become acutely aware of the physical discomforts of their surroundings. They should be given numerous opportunities for taking notes and for equally essential sketching and photography.

This type of fieldwork frequently involves visits to a number of points, perhaps half a dozen in a day, and the impression may tend to be given that these are separate little studies, totally unrelated one to the other. It is therefore highly desirable that such points should be most carefully selected, and connected by linking commentary made between stops. Care should, however, be exercised to see that a linking commentary, especially if it is given in a bus through a public-address system, does not become an excuse for a continuous monologue; nothing kills interest quicker than a steady stream of disembodied words. The "look and see" type of field work is infinitely more successful if conducted on foot with time to digest what is seen; the bus at best is a means of getting from place to place, not a substitute for feet on the ground. During such field days the participants should be encouraged to do some work of their own, not be expected just to sit and receive. A socratic approach can be adopted and an attempt can be made to elucidate the facts and ideas one wishes to impart from the pupils themselves. A useful device in this type of field work is the backward view; having shown a class a particular piece of terrain from one viewpoint, the same piece of terrain may be shown to them later from a different angle, when it is possible to recapitulate the points made at the first stopping point.

The methods and virtues of field note-taking are two most important considerations which are frequently neglected. Many dislike taking notes in the field. From time to time one has the feeling that observations are so obvious and ideas so striking as to be unforgettable. But as new experiences are piled one on top of another, early ones are soon forgotten no matter how dramatic and original they were. On the other hand, the continuous taking of notes at the expense of looking and seeing is to be decried. There should be no need to make notes more than on three or four occasions during the day. Indeed it may be worth experimenting by getting pupils to make no notes at all during a day in the field but to sit down in the evening and make their notes then; note-taking in the field can then be confined to sketching of both maps and views. How commonly are pages of nearly illegible so-called field notes gained at the expense of the pupil's sight and comprehension of what he has been brought to see!

FIELDWORK PROJECTS

Above all project work in the field needs to be planned well in advance. There is a tendency to make such projects too long and too difficult: they must be within the intellectual capacity of the pupils and capable of being completed within the time available. There is nothing more demoralizing than to be required to do a piece of work clearly beyond one's technical competence or incapable of being completed within the time available. In this type of work it is more likely that the ability and speed of a group as a whole will be closer to that of the weakest and slowest members. A great deal depends upon adequate briefing and this requires that the teacher shall have himself done the project before he asks the pupil to do it, and to have asked himself in particular what snags are likely to be encountered by his pupils.

The amount of factual data which is collectively produced during such project work can be enormous, but the mere acquisition of data is of little value unless it can be processed and put into a meaningful form for inclusion in a field notebook. It must have a purpose which is made clear to the participants from the outset and not be simply work for work's sake. It is much better to limit the scope of a project and get worthwhile data which can be adequately processed than to acquire masses of material which pupils have no hope of being able to handle. Short, sharp projects, well circumscribed and completely worked out, are infinitely better than long, amorphous ones in which pupils tend to lose themselves and to lose heart.

RESEARCH DATA FROM THE FIELD

In the past some of the most valuable research work in the field sciences has been carried out by amateurs with a deep interest in their local environments. In England, Gilbert White, a country parson, set the pattern for such local studies with his *Natural History of Selborne*. In the nineteenth and twentieth centuries in Britain this type of work has characterized local natural history, geological, and archaeological societies. Geography as such has been less well catered for, although the school teacher and his pupils, particularly those in the sixth form, have tremendous opportunities to contribute significantly to geographical research. The latter-day Gilbert Whites could be, and indeed sometimes are, the local school teachers who, together with their senior pupils, are ideally placed to contribute significantly to geographical research. The pupils live scattered over the whole of the school's catchment area, their collective knowledge about their environment is enormous, and they can be trained to acquire much useful data. Even the mere collection of data, if done under controlled conditions, is of educational value, for it soon reveals the inherent weaknesses which underlie statistics such as mean annual temperature figures. The geography master can act as originator and co-ordinator of many potentially useful projects. A few suggestions follow, taken from the field of physical geography, although equivalents could be listed for the human aspects of geography.

At the heart of some geographical fieldwork is the need to understand the nature of the underlying geology. In Britain, such an understanding may be obtained in part from the relevant geological maps on a scale of 1 in. to 1 mile, or the 6 in. to 1 mile field sheets of the Geological Survey, but a much greater appreciation comes when these maps are related to exposures of the rocks in the field. Exposures of a semi-permanent character are fewer now than they were in the nineteenth century, but temporary excavations such as house foundations, sewer and pipeline trenches, and road cuttings abound. They are soon lost and it is desperately important that a record should be made of what is revealed in them. If this sounds a tall order for a geography teacher, considerable help is to be obtained from elementary field manuals in geology (e.g. Himus and Sweeting, 1955). Particular note should be made of the top part of such sections where, in the weathering zone and the soil, evidence of geomorphological processes such as frost shattering may be found. Such straightforward things as angle of dip and joint spacing are readily observed and measured and their relationship to the surface of the ground examined. Such temporary sections also give a view of the soil profile but augering to a soil auger's depth will usually reveal important features of the soil profile in areas where there are no temporary sections. If a more comprehensive view is required it is always possible to dig soil pits. An examination and description of the principal horizons in a soil profile can then readily be made. It should also be possible in a school to make some elementary analyses of soils. For example, there may be scope here for the measurement of pH values, and for co-operation with the chemistry master on the interpretation of the results. The *Soil Survey Memoirs* are invaluable guides to the types of things to look for (see Chapter 5, above).

Landforms are readily analysed. Elementary levelling with an abney level, for instance, is a perfectly adequate method of measuring slope profiles. The mapping of breaks and changes in slope as described by Waters (1958; see also Chapter 2, above) is within the capabilities of senior pupils who can be encouraged to make geomorphological maps using a variety of symbols for such features as scarp slopes, terraces, fans, and knickpoints. A variety of such symbols are described by Dury (1952), which are as applicable to field-made maps as to any others. It should also be possible for a school group to attempt some measurements of rates of erosion in many different local environments (see Chapter 2, above). There are many other possibilities, too, for original work on the nature of superficial deposits, such as analyses of the sizes and shapes of pebbles in the bed load of a river or in a glacial deposit.

One of the most neglected fields of study in physical geography is that of hydrology. As Chapter 6 above shows, there is a desperate lack of elementary hydrological data which schools could help to fill. The measurement of stream velocity and discharge is possible without sophisticated instrumentation, and the keeping of regular records of streamflow of local streams would be a very useful contribution to knowledge. Measurement of the nature of a river's load—its bed load, the load carried in suspension and in solution—is susceptible to analysis by relatively elementary means. Studies of such phenomena in the field can be related to map study of the area and morphometric properties of the relevant drainage basin as determined from maps. The study of contemporary geomorphological processes is a lively theme in current geomorphological research and school groups could make a significant contribution to knowledge by keeping records of what is happening in their school area.

The school weather station is now a well-established feature of the life in many schools and should be part of every school's equipment. The keeping of records is good training in scientific discipline for pupils and provides invaluable teaching material (see Chapter 3, above). But school groups have it within their powers to investigate climatic phenomena over a wider area than is covered by a single weather station. It would be relatively simple to organize a study of visibility in the vicinity of the school if pupils were to be asked to record how far they could see at a specified time, before they left their homes for school on particular mornings, through the year, by reference to some pre-established scale of distance. Such data could then be mapped and isopleths drawn to show, say, the distribution of fog on particular mornings of the year. An elementary but nevertheless valuable contribution to air-pollution studies could be made by exposing to the atmosphere for a specified time, previously weighed Petri dishes in, say, the gardens of pupils' homes, which could then be carefully weighed on a physical balance to obtain the weight of sediment which has settled on the dish. Or again the dates of first flowerings of specified plants could also be recorded for many different places in the school catchment area and isoline maps constructed showing dates of first flowerings. Such phenological studies represent a concise summary of the effects of climatic conditions in the local environment. What can be achieved by schools in co-operation in the climatological field has been demonstrated by T. J. Chandler in the London Climatological Survey. Temperature traverses across a valley or through a town to measure the effect of relief and urban environment (Chandler, 1965), are other possible fruitful fields of endeavour.

In the field of vegetation studies, if pupils could be taught to recognize the dominant species in an area, then the identification and mapping of the principal vegetation communities on the basis of dominant species should not be beyond sixth-form capabilities (see Chapter 4, above). The use of quadrats and transects to demonstrate the areal pattern of vegetation is quite feasible. School groups are well placed to trace changes in vegetation through time. There is considerable interest and value to be obtained from a sequence of photographs taken of the same locality under the same lighting conditions at different seasons of the year and over a period of years. This type of time-lapse photography is also applicable to some geomorphological changes such as soil erosion. A school could, for instance, adopt a piece of heathland or a gully and photograph its development through time, thus building up a potentially useful body of data.

In all such investigations it is important that the data should be measured in some way, or, in contemporary jargon, that it should be quantified. It is then often susceptible to statistical treatment and could serve as a basis for the classroom teaching of elementary statistics. A further step would be to introduce pupils to the use of punched cards for the recording of field data—it is then an easy step to the concept of data processing and the understanding of the principles of the electronic computer. Clearly, not all these suggestions could be followed up at any one time by any one school, but a few small projects set up in a school area would be of inestimable value in the geographical education of pupils. One might also hope, through contacts with local interested persons such as engineers and surveyors, to interest pupils in the applications of their knowledge of their physical environment.

THE INTEGRATION OF PHYSICAL GEOGRAPHY

The study of geography suffers, and perhaps always will suffer, from centrifugal tendencies. It is extremely difficult, some would say impossible, to comprehend its totality. As a consequence the systematic branches of geography have been conceived, some concerned with physical, others human aspects of the earth's surface, but all capable of standing in their own rights as separate intellectual disciplines. Current methods of teaching physical geography in both schools and universities recognize this fact and usually involve the teaching of separate lessons and courses in geomorphology, climatology, biogeography, etc. It is, however, very necessary for the continued existence of geography that we should try to develop an integrated approach in its teaching, not least in physical geography. The centrifugal tendencies of the classroom must be counteracted, and one way in which this may be done is by a deliberate attempt to provide integrated studies in the field.

In the field it is very difficult to ignore the interrelationships between the various aspects of physical geography, difficult to study the geomorphology of a valley without taking into account its soils, streams, and vegetation. But it is equally difficult to study these interrelationships by focusing attention directly upon them. They are best studied indirectly through one of the physical elements themselves. At an elementary level climatically-based studies of physical geography are often used. Alternatively, the study of physical geography may be approached through a study of plants and animals. This method has been highly developed by ecologists, under the guise of the ecosystem. But as with the climatic approach, this is again essentially one best suited to the classroom. If water is taken as the core of physical geography, then the hydrological cycle provides a useful vehicle for integration. The movement of water from the atmosphere on to the land, to the oceans, and back into the atmosphere, linked with an examination of its use by plants, animals, and men, is a logical and easily comprehensible systems approach. But again it is perhaps better suited to the classroom rather than field studies.

The fourth possibility is to approach the study of an integrated physical geography through the surface of the land itself. Such an approach in the field could usefully start with a study of an individual slope, for example, that from the crest of a scarp down to its foot, or that of a valley side from the crest of an interfluve to the stream in the bottom of the valley. Measurements of all aspects of the physical environment can be made along such a profile; altitudes, slope angles, slope aspect, the nature of the underlying geology, changes in soil type, the presence or absence of water at different points on the slope, vegetational differences, and, if time and knowledge permit, differences in animal life. This integrated study can be carried further into the field of human geography and an analysis can be made of land use, communications, and settlement as they are developed on that particular slope. In soil studies such a sequence is referred to as a "catena", and we might usefully call the sequence of geographical changes along a slope profile a "geographical catena".

The study of a series of individual catenas can be integrated into a transect. Transects can be studied at a variety of scales. At a general scale, it is possible to make a detailed study of the physical geography of, say, scarp and vale topography extending over a distance of several miles along a line at right angles to the grain of the country. More detailed transects are possible and may be of value, such as one across a line of sand dunes and slacks of successively greater age from the present coast to an old cliff line.

The catena and transect are best summarized in the form of a transect diagram on which slope, soil, drainage, etc., can be described and recorded with reference to the drawn profile which forms the basis of the diagram (e.g. Mead and Brown, 1962).

The study of catenas and transects can be developed into studies of areas with the help of the "site" concept as developed by Bourne (1931) in forestry studies and by Linton (1951) in geomorphology. A site is the smallest indivisible morphological unit of land shape having a unitary slope angle and orientation and is bounded by breaks, changes, and inflections of slope (Savigear, 1965; see also Chapter 2, above). The up-slope convexity in a scarp slope catena is such a unit and the bounding changes of slope which separate it from the scarp face below and the interfluve above are entirely capable of delimitation on, say, a 6 in. to 1 mile map by a sixth-form pupil in the field. Each such morphological unit or site is susceptible to analysis with respect to all aspects of its physical geography in a manner similar to the suggested treatment of the catena and transect. As area is a variable the analysis of the distribution of sites as mapped in the field is recommended. Such a method of quantitative analysis is outlined by Gregory and Brown (1966).

Assemblages of sites make up regions such as a single valley or a single plateau, which were called "stows" by Unstead (1933). These in turn may be aggregated to form "tracts" of wider extent, such as whole cuestas or strike vales. By building up a hierarchy from catenas, transects, sites, to stows and tracts from field observation, it is possible to move into the classroom and to study larger areal units or regions on large-scale maps. In this way, not only may the component parts of physical geography be integrated in the field but there is an opportunity to establish links between what is visible to the pupil in the field and what he reads in his textbook in the classroom.

REFERENCES AND FURTHER READING

INTRODUCTION

The literature on approaches to fieldwork is not a particularly rich one. Reference has been made in the text to a number of sources which are helpful over particular problems. Such are:

CURTIN, W. G. and LANE, R. F. (1955) *Concise Practical Surveying*, English U.P., London.
DURY, G. H. (1952) *Map Interpretation*, Pitman, London.
HIMUS, G. G. and SWEETING, G. S. (1955) *The Elements of Field Geology*, U. Tutorial P., London.
KING, C. A. M. (1966) *Techniques in Geomorphology*, Arnold, London.

See also:
Commission on College Geography (1968) Field training in geography, *Association of American Geographers Technical Paper*, **1.**

Guides to field excursions are illustrated in:

CLAYTON, K. M. (Ed.) (1964) *Guide to London Excursions*, London School of Economics.
Geologists' Association Guides, Benham, Colchester.
WOOLDRIDGE, S. W. and HUTCHINGS, G. E. (1957) *London's Countryside*, Methuen, London.

Much concerning the nature and practice of fieldwork in Britain is to be gained by attendance at the courses arranged at field-study centres by the Field Studies Council whose headquarters are at 9 Devereaux Court, London, W.C.2.

RESEARCH DATA FROM THE FIELD

Geomorphological mapping techniques are described in:

SAVIGEAR, R. A. G. (1965) A technique of morphological mapping, *Annals of the Association of American Geographers*, **55,** 514–38.
WATERS, R. S. (1958) Morphological mapping, *Geography,* **43,** 10–17.

The valuable work of schools' climatological investigations is illustrated in:

CARR-GREGG, R. H. C. (1961) Meteorology and climatology in schools, the Lea Valley Climatological Survey, *Geography*, **46,** 307–14.
CHANDLER, T. J. (1965) *The Climate of London*, Hutchinson, London.

Many of the things which teachers and their pupils can do in the way of measurement in the field are outlined in:

RIGG, J. B. (1968) *A Textbook of Environmental Study*, Constable, London.

THE INTEGRATION OF PHYSICAL GEOGRAPHY

Concepts of catena, transect, site, morphological unit, etc., are to be found in:

BOURNE, R. (1931) Regional survey, *Oxford Forestry Memoirs*, **13,** 16–18.
GREGORY, K. J. and BROWN, E. H. (1966) Data processing and the study of land form, *Annals of Geomorphology*, **10,** 237–63.
LINTON, D. L. (1951) The delimitation of morphological regions, in STAMP, L. D. and WOOLDRIDGE, S. W. (Eds.) *London Essays in Geography*, Methuen, London, pp. 199–217.
MEAD, W. R. and BROWN, E. H. (1962) *The United States and Canada*, Hutchinson, London.
UNSTEAD, J. F. (1933) A system of regional geography, *Geography*, **18,** 175–87.

PART II

HUMAN GEOGRAPHY

All the mighty world
Of eye and ear, both what they half create
And what perceive.

WORDSWORTH, *Lines composed a few miles
above Tintern Abbey*

CHAPTER 8

QUANTIFICATION AND THE DEVELOPMENT OF THEORY IN HUMAN GEOGRAPHY

W. J. CAMPBELL and P. A. WOOD

THE SEARCH FOR ORDER AND THEORY IN HUMAN GEOGRAPHY

The most persistent modern trends in human geography are based upon the formalization of ideas into theories and models and involve the development of procedures of investigation and analysis employing mathematics and statistics. Such trends are indicative of the growing use of scientific methods of inquiry in the investigation of geographical problems. Essentially, this new approach provides methods of deductive reasoning and inductive inquiry, which allow the truth of our ideas about certain aspects of the world to be tested rigorously in the light of factual evidence. The broader aim underlying these procedures is to connect different ideas together and to build them into coherent explanations or theories. The adoption by geographers of these widely accepted methods of description and explanation is bringing the discipline more fully into the mainstream of scientific investigation, which is concerned with the search for order in a complex world.

One reason why these developments may seem new to geography is that the subject has a strongly empirical tradition which stems, at least in part, from the evident difficulty and challenge of gathering information on a very wide range of phenomena, at many scales of study and under ever-changing conditions. In practice, however, this process of observation can never proceed without some, often unconscious, judgement about the relative significance of the facts collected. As Burton put it, "the moment that a geographer begins to describe an area . . . he becomes selective (for it is impossible to describe everything), and in the very act of selection demonstrates a conscious or unconscious theory or hypothesis concerning what is significant" (Burton, 1963, in Berry and Marble, 1968, p. 18). All forms of productive inquiry, even those traditional to geography, must therefore embody a theory, an idea, or even a prejudice for which support is being sought. Highly generalized, often loosely formulated theories or "metaphors" have long been present in the subject. Jefferson's "law of the primate city" provides a well-known and long-established example in urban geography. Nevertheless, "pregnant as such metaphors are with concepts, ideas and generalizations, there comes a time when the form in which such metaphors are cast seems to hinder objective judgement. The wish to be objective and scientific is somehow frustrated" (Harvey, 1967, p. 551).

It is precisely this frustration that scientific methodology seeks to overcome, for it provides guidelines by which a specific idea can be investigated and critically accepted or rejected. It also demands that the investigator explicitly recognizes the underlying assumptions of his methods and the limitations which these impose on his conclusions. Loosely structured ideas, based upon intuitive judgement, may thus be replaced by statements of much greater precision, which have been or can be objectively tested. Elements of this approach have been present in the past, but geographers in general have been unwilling or unable to present a coherent end-product in the form of a well-defined theory which explains patterns and processes of geographical interest. More effective progress is possible when such theories are established and become accepted as the framework for empirical studies. An example is central-place theory, which is the basis for most modern studies of the size, nature, and spacing of market and service activities (see Chapter 18, below).

A theory may therefore be defined as a set of statements about relationships that have been established with some degree of confidence and, as a result, provides a foundation for future investigation. A theory must state explicitly the assumptions upon which it is based, and these may be abstract or greatly simplified to eliminate the effect of variables that are not of central interest. It should identify the processes which underlie the relationships it seeks to explain, and so allow extension of this explanation to predictions of occurrences that have not already been observed but which might be expected. Predictive value is a crucial test of a good theory but this quality can never be expected to extend outside the range of conditions and circumstances embodied in its formulation. A theory can be tested, verified, rejected, or modified according to the rules of scientific investigation, and in the long term adequate explanation is impossible without theory (Hempel, 1966).

The relationships that theories seek to explain are applicable at specific levels of generalization. Thus even human behaviour, with all its individual variations, can be described accurately, provided explanation is made at an appropriate level of generality. The activities of groups of people over space and time have, in fact, been quite adequately described by geographers. For example, diffusion models based on the rules of probability have successfully represented the spread of ideas and innovations (Hägerstrand, 1967), while movements of many types between areas have been effectively generalized by gravity and migration models (Huff, in Berry, 1967; Morrill, 1965). Refinement of these models continues to be a matter of active concern, but it is generally agreed that such problems can be dealt with successfully.

In summary, the major impact of a scientific approach in geography has been twofold. First, it has given the subject new and more effective methods for analysing complex patterns and relationships and, provided they are used with circumspection, these methods can often demonstrate the otherwise unrecognized existence of order in geographical distributions. Second, it has emphasized the search for theoretical explanation, with a consequent redefinition and refinement of the fundamental concepts and ideas on which geographical inquiry is based.

SPATIAL ANALYSIS AS A FOCUS FOR HUMAN GEOGRAPHY

In his review of the nature of geography, Hartshorne indicated that the purpose of the subject has been "to provide accurate, orderly and rational description and interpretation of the variable character of the earth's surface" (Hartshorne, 1959, p. 21). This task incorporates many concepts which can be attributed to other disciplines. If these are laid aside, there remains a group of ideas concerned specifically with spatial variations and relationships. These ideas deal with the nature and effects of distance, location, pattern, and shape, which are derived from the basic elements of point, line, and area. Spatial analysis attempts to identify, describe, and interpret regularities between these elements, relating such regularities to the processes which produce them. This is not entirely a new emphasis, but one that has received impetus and a coherent structure from the desire to construct distinctively geographical theory.

Among the many examples of this emphasis in modern work, a few may be mentioned. Transport networks can be analysed as a series of points and connecting links (see Chapter 16, below). Many distributions, such as those of settlements, can be viewed as point patterns (Dacey, 1962; King, 1962; see also Chapter 12, below), and these patterns classified according to their geometry or the types of processes that may have produced them, for example uniform, random, competitive or contagious processes (Harvey, 1967, p. 554). Area problems have focused on the methods of boundary definition (Yeates, 1963), on the grouping of small areas into larger regions (Berry, 1960), or on the identification of surfaces and gradients representing, for example, densities of population or intensity of economic activity (see Chapters 18 and 20, below).

In spatial analysis the search for theory and order receives support from two directions. First, there is the regularity that might be expected from the well-established mathematical study of space, i.e. geometry. This provides a nomenclature, language, and set of theorems which may be applied to points and lines on any surface, including that of the earth (Bunge, 1966). The second source of encouragement in seeking geographical order comes from the world itself. The effects of distance and location consistently emerge from many studies concerned with such wide-ranging topics as farming types, settlement patterns, transport flows, industrial location, and the functional zones of cities.

The manner in which the scale of study may affect the type and degree of order found in spatial analysis has attracted much attention. Recognizable patterns on a local scale may lose their significance at a regional level. Conversely, apparently meaningless patterns in a detailed study may become explicable simply by looking at them in a much wider context. The degree of identifiable regularity and generalization possible in human behaviour certainly changes with scale and with the number of observations considered. These points can be illustrated by considering such topics as shopping and migration behaviour, or social organization within cities (for example, see Chapter 9, below). The idea of scale as a factor in human organization and interaction is at least as important as the concept of scale in the linear sense and, of necessity, attention must be given to the relationship between the two.

One of the most difficult problems in developing theory is the need to adopt simplifying assumptions. In spatial theory attempts to isolate the effects of distance and location often involve this practice, which might appear to remove such studies from the realm of observed reality. Such simplifications are, however, the basis for all analytical thought and the value of the resulting theories lies in the insights they provide into certain aspects of reality, rather than into the whole of it.

In spatial analysis, the best known simplifying assumption is the flat, uniform (or isotropic) plain, "the elementary, abstract geographical space that has no difference from place to place or in one direction to another" (Nystuen, 1963, in Berry and Marble, 1968, p. 37). A comparable concept in economics, which is also incorporated into classical location theory, is the behaviour of "economic man", who has perfect knowledge of events and perfect ability to act in accordance with the single motivation of maximizing profits. This imaginative idea has proved to be crucial in the progress of economics as a theoretical discipline. The argument for such simplifying assumptions in spatial studies was made as early as the 1820's by von Thünen, who explained in the preface to the second edition of his major work, that "they are a necessary part of my argument, allowing me to establish the operation of a certain factor, a factor whose operation we see but dimly in reality, where it is in incessant conflict with others of its kind" (von Thünen, 1966, p. xxii). In the development of location theory, the isotropic plain has been an invaluable assumption which has allowed the distinctive properties of distance, location, and the patterns produced by various processes to emerge clearly, with other complicating effects held constant.

When the simplified situation is well understood, some of the initial assumptions can be relaxed and the effects introduced of physical barriers, unequal transport developments, uneven resource distributions, or other complications. Such complications can be analysed systematically in terms of the modifications that they cause to the influence of distance through, for instance, varying ease of travel and costs of movement. Distance measured in conventional linear terms may be the least interesting way of assessing spatial relationships (Hägerstrand, 1957; Bunge, 1966). One of the important skills of the geographer is his ability to think in terms of the many transformations of distance and location that have relevance to human behaviour. A very promising, but as yet scarcely understood transformation is that which takes account of the different ways in which man perceives distance and space around him and reacts to it (see Chapter 11, below).

The rich possibilities of thinking about spatial problems in theoretical terms, and the insights this can produce, provide a common thread running through much contemporary human geography. Attention is now turned to one of the fundamental requirements of any good theory, which is that its ideas can be tested and related to empirical observations.

THE RÔLE OF QUANTITATIVE METHODS IN TESTING THEORY

The significant increase in the number of geographical studies using quantitative methods of analysis during the decade 1950–1960 has been outlined elsewhere (Burton, 1963), and it is now hardly necessary to argue the need for their use. What should be emphasized, however, is that the distinction often made between "quantitative" and "qualitative" analysis is not very useful. The widespread employment of these methods in science is only a symptom of something more fundamental: the true value of mathematical and statistical methods lies in the rôle that they play in the testing and verifying of ideas established, however tentatively, within a theoretical framework. As Burton stated, "it is not certain that the early quantifiers were consciously motivated to develop theory, but it is now clear to geographers that quantification is inextricably intertwined with theory" (Burton, 1963, in Berry and Marble, 1968, p. 19).

An important consequence of this for the teaching of the subject is that the understanding of modern developments does not depend, in the first place, upon any detailed working knowledge of mathematics or statistics. It is more important to have an appreciation of how ideas are developed in science, in the form of groups of hypotheses or models. Only when the logical basis of a particular set of ideas is grasped does the need follow to translate them into forms that are susceptible to testing against factual evidence. Virtually all statistical methods, with the exception of descriptive statistics (means, standard deviations, etc.), are designed to test ideas or hypotheses. For example, they may test whether a relationship exists between variables (e.g. correlation) or whether two groups differ markedly in the properties they possess in common (e.g. analysis of variance). This dependence of statistical methods upon the correct formulation of ideas or hypotheses becomes particularly important when dealing with more complex methods of multivariate analysis (e.g. multiple regression, factor analysis). These techniques are designed to search for relationships of a specific type between many variables. Only rarely can original insights emerge from the largely uncritical and unstructured statistical analysis of data.

Having said this, it should be added that the act of understanding a particular method may illustrate the way in which questions or ideas must be cast for the purpose of objective analysis. A knowledge of basic mathematics is certainly useful in geography today, and perhaps essential at the research level, since this skill allows the development and testing of theoretical ideas through the construction of models.

This wider view of quantification leads necessarily to an understanding of the rôle of models in geography. The term "model" is often used in different senses, but in the context of scientific method it has a specific meaning. A model is the formal presentation of a theory, often in mathematical form, and is constructed for the specific purpose of enabling the relationships embodied in the theory to be tested. Individual theories only attempt to explain certain limited phenomena, so that models designed to test them represent only the relationships of immediate interest. Such is the complexity of organization of some human activities, particularly as they change over time, that an adequate summary of all their aspects cannot be contained within any single model, with its necessary simplifying assumptions. As a result, several interrelated models may have to be used. This need for linked models is essentially similar to the idea of a system, which has been adopted in several branches of geography to describe complex, functionally related features or sets of objects (for example, see Chapters 4 and 13).

The arguments for the use of models have been widely reviewed and discussed (Chorley and Haggett, 1967) as have classifications of models and their uses (Chorley, 1964). Possibly the most important current distinctions are those between deterministic and stochastic models, and also whether these are static or dynamic (Harvey, 1967). It is indicative of the rapid progress of geography in this direction, at least in certain fields, that the most active and interesting research effort is now concentrated on the very difficult stochastic–dynamic models. In these, random or chance factors known to operate in human activities are incorporated into the regularities underlying the evolution of spatial patterns (e.g. models of the diffusion and spread of innovations and colonists). The rôle of the electronic computer in this work has been a crucial one, not only enabling large amounts of data to be rapidly analysed but, more important, permitting the manipulation and testing of more complex models, which make heavy and repetitive computational demands.

The revolutionary phase of quantification in geography is now ended. This is emphasized by the fact that a few geographers are beginning to make original contributions to the specific problems of spatial statistics (arising, for example, from differing sizes and shapes of areal statistical units and from the interdependence of places in spatial proximity to each other). Certainly, the use of quantitative methods is now much more selective and critical than it was in the 1950's, as several important collections of examples of their use have recently shown (Garrison and Marble, 1967; Berry and Marble, 1968; Smith *et al.*, 1968). These works, and current geographical publications in general, indicate that they are now accepted as basic tools and techniques. It is vital that this be the case and that attention is concentrated on the concepts and problems of spatial analysis and the methods of constructing theory. Lack of ideas and the ability to formulate them effectively will in the long run be profoundly more damaging to the intellectual progress of the subject than a slightly less than perfect ability to handle statistical techniques.

CONCLUSION

This chapter has discussed some highly significant innovations in human geography and there seems little doubt that at least some of the concepts that we have mentioned are having or will have important repercussions throughout the subject. Two further points need to be made, however, in order to put these developments into perspective.

First, the adoption of a theoretical approach has clearly gone further in some branches of human geography than in others. As a result, there is a danger of exaggerating the relatively modest progress in some fields when comparison is made with other parts of the subject where the pace of change has been slower. Economic, urban, and population geography have been affected most profoundly, allied with developments in closely related social sciences. Cultural and historical geography, however, seem less ready for the development of theory, although already their subject matter poses some questions which cast doubts upon the general applicability of theories being developed in other systematic branches of study. For example, they raise the whole question of comparisons between different cultures and modes of perception.

Second, it is clear that the general approach to the study of human geography we have described is somewhat idealized. Many of its facets are not fully understood or accepted and they cannot all be implemented with equal ease. A great deal of attention has been paid to the application of standard statistical procedures, perhaps at the expense of real advances in the derivation of theory. It can hardly be expected, therefore, that the changes implied by this approach will be revolutionary in nature. Analogy with other disciplines suggests that it is quite normal for different modes of thought or traditions of inquiry to co-exist for long periods of evolutionary transition. Rarely, as change occurs, are all links with earlier traditions completely severed and often ideas or viewpoints which have been long discarded are taken up again as new methods or information renew their significance. For example, the foundations of theory in spatial analysis lie in the classical economics of the nineteenth century, and the types of generalization now advanced to describe geographical behaviour were once rejected as being too deterministic.

All that is demanded by any potentially fruitful line of approach and inquiry is that students who are intellectually committed to the discipline should be willing to judge the significance of such an approach and to contribute to its development in a constructure manner whenever possible.

REFERENCES AND FURTHER READING

THE SEARCH FOR ORDER AND THEORY IN HUMAN GEOGRAPHY

Several additional references are given which deal specifically with the philosophy and methods of science or with their applications in geography.)

BERRY, B. J. L. (1967) *Geography of Market Centers and Retail Distribution*, Prentice-Hall, Englewood Cliffs.

BURTON, I. (1963) The quantitative revolution and theoretical geography, *Canadian Geographer*, **7,** 151–62.

HÄGERSTRAND, T. (1967) On Monte Carlo simulation of diffusion, in *Quantitative Geography: Part 1, Northwestern University Studies in Geography*, **13,** 1–31.

HARVEY, D. (1967) Models of the evolution of spatial patterns in human geography, in CHORLEY, R. J. and HAGGETT, P. (Eds.) *Models in Geography*, Methuen, London, pp. 549–608.

HARVEY, D. (forthcoming) *Explanation in Geography*, Arnold, London.

HEMPEL, C. G. (1966) *Philosophy of Natural Science*, Prentice-Hall, Englewood Cliffs.

MORRILL, R. L. (1965) Migration and the spread and growth of urban settlement, *Lund Studies in Geography, series B*, **26.**

National Science Foundation (1965) *The Science of Geography*, Washington.

SPATIAL ANALYSIS AS A FOCUS FOR HUMAN GEOGRAPHY

In addition to the limited number of studies listed here, mention has been made above to issues discussed in other chapters, and the reader is referred to the works cited in these.

BERRY, B. J. L. (1960) An inductive approach to the regionalization of economic development, *University of Chicago, Department of Geography Research Paper*, **62,** 78–107.

BUNGE, W. (1966) Theoretical Geography, *Lund Studies in Geography, series C,* **1,** revised edn.

DACEY, M. F. (1962) Analysis of central place and point patterns by a nearest-neighbour method, *Lund Studies in Geography, series B,* **24,** 55–75.

HÄGERSTRAND, T. (1957) Migration and area: survey of a sample of Swedish migration fields and hypothetical considerations on their genesis, *Lund Studies in Geography, series B,* **13,** 27–158.

HARTSHORNE, R. (1959) *Perspective on the Nature of Geography,* Association of American Geographers, Chicago.

KING, L. J. (1962) A quantitative expression of the pattern of urban settlements in selected areas of the United States, *Tijdschrift voor Economische en Sociale Geografie,* **53,** 1–7.

NYSTUEN, J. D. (1963) Identification of some fundamental spatial concepts, in BERRY, B. J. L. and MARBLE, D. F. (1968) *Spatial Analysis: a reader in statistical geography,* Prentice-Hall, Englewood Cliffs, pp. 35–41.

VON THÜNEN, J. H. (1966) *Von Thünen's Isolated State,* edited by P. Hall and translated by C. M. Wartenberg, Pergamon, Oxford.

YEATES, M. H. (1963) Hinterland delimitation: a distance minimizing approach, *Professional Geographer,* **15,** 7–10.

THE RÔLE OF QUANTITATIVE METHODS IN TESTING THEORY

BERRY, B. J. L. and MARBLE, D. F. (Eds.) (1968) *Spatial Analysis: a reader in statistical geography,* Prentice-Hall, Englewood Cliffs.

CHORLEY, R. J. (1964) Geography and analog theory, *Annals of the Association of American Geographers,* **54,** 127–37.

CHORLEY, R. J. and HAGGETT, P. (Eds.) (1967) *Models in Geography,* Methuen, London.

GARRISON, W. L. and MARBLE, D. F. (Eds.) (1968) Quantitative Geography; Part 1: economic and cultural topics, *Northwestern University Studies in Geography,* **13.**

SMITH, R. H. T., TAAFFE, E. J. and KING, L. J. (Eds.) (1968) *Readings in Economic Geography: the location of economic activity,* Rand McNally, Chicago.

YEATES, M. H. (1968) *An Introduction to Quantitative Analysis in Economic Geography,* McGraw-Hill, New York.

These additional references provide a simple introduction to statistics:

GREGORY, S. (1968) *Statistical Methods and the Geographer,* 2nd edn., Longmans, London.

HAYSLETT, H. T. (1967) *Statistics Made Simple,* Allen, London.

REICHMANN, W. J. (1964) *Use and Abuse of Statistics,* Penguin, Harmondsworth.

CHAPTER 9

POPULATION GEOGRAPHY

J. W. WEBB

UNTIL quite recently the place of population studies in geography was uncertain. While geographical inquiry was still focused on landscapes and regions, authors did not always consider it important to treat population as a geographical variable. Although Vidal de la Blache (1922) and others conceived of human geography as dealing with the variable distribution of population on the earth, some geographers hardly mentioned population in the list of phenomena to be integrated into the study of a region.

By the 1950's there were indications of a growing branch of geography distinct enough to have its own body of methodological and substantive literature (Trewartha, 1953; Beaujeu-Garnier, 1956, 1958, and 1966). Since then the ground has solidified with a general primer (Clarke, 1966), a programmatic statement (Zelinsky, 1966), an extensive bibliography (Zelinsky, 1962), and two short discussions relating population studies to geography (Wrigley, 1965 and 1967). At the same time geography itself has undergone, and is still undergoing, a shift in the general procedures by which it achieves its intellectual aims. This brief survey relates recent developments in the subject to changes in geography as a whole and discusses some other issues which are relevant to the teaching of population geography.

RECENT DEVELOPMENTS IN POPULATION GEOGRAPHY

In the regional approach, which dominated geographical teaching a generation ago, the study of population seemed to be vaguely important but in the last analysis was usually relegated to a minor position. One proceeded from the "physical basis" to the description of economic (and sometimes social) phenomena, using physical patterns as the primary explanation of human activity. There were few references to relative location, and a preoccupation with physical site underpinned the subject. Population was overlaid on the patterns of topography, water supply, climate, and vegetation. The visual and mental relationships that emerged were perceived intuitively but described with assurance.

The rigidities of this classical procedure had its advantages; in the hands of a master the geography of a place or region could be a beautifully integrated descriptive study. But often, and especially in the classroom, the procedure was mechanical and the explanatory connection between society and environment was diagnosed by rule of thumb. Those without the necessary persuasive literary gifts resorted to unrelated descriptions of separate phenomena, each under its own heading. Like other aspects of man, population was a passive element in the discussion, often inadequately dealt with by a numerical accounting reminiscent of a statistical yearbook.

Systematic or topical studies were subservient to the regional synthesis, but in the last 40 years the various specializations dealing with man have separated out and such branches as urban, industrial, and population geography have developed their own momentum and achieved an existence independent of the regional syntheses they were supposed to support. During the same time, most research workers abandoned the possibility of producing all-encompassing regional studies. In their place systematic specializations were developed in which the internal nature of the subject under consideration provided its own rationale. Systematic studies are in fact just that— studies of systems which attempt a description and explanation of the components in a particular system, including the links between the components. Of the external forces and factors which make each system subject to change, the effects of location are of most interest to geographers. It is the study of location, whether conceived of as the internal nature of places or as the relations between them, that bind⁻ geographers together, whatever their individual systematic interests.

Demographers and sociologists can treat populations as devoid of spatial context; but geographers who treat population as some arithmetic expression, unrelated to place or space, do so at the peril of losing contact with their discipline. Zelinsky (1962, p. vi), in his definition of the subject, put population geography firmly within the geographical tradition: it "studies the ways in which spatial differences in population are related to spatial differences in the aggregate nature of places". It deals with those population phenomena which help to form, and are reacted on by, the character of places. One could define almost all other branches of geography using similar words.

Some population geographers argue for a much greater recognition of the significance of population studies by geographers generally. Hooson (1960) thought that, since geography has become increasingly man-oriented, the explanation of the areal distribution of man himself has become the central geographical problem. Wrigley (1965 and 1967) argued that it is no longer necessary to view population distribution as the result of economic and other forces; rather, population distribution can be examined in its own right as a cause and not only as an effect. Population geography is more than an autonomous branch of the subject, it is the touchstone of relevance for geography as a whole. Population distribution can explain economic activity, and its consideration should be the start and end of analysis in human geography.

Perhaps this claims too much for population geography. In the past there has been little success for geographers who have referred all questions to a single category of phenomena for explanation. For the time being at least, locational analysis would appear to be the common ground of geography, and location is not a category or group of phenomena but a condition of all phenomena.

SOME ISSUES IN POPULATION GEOGRAPHY

Scale

The concept of scale is so well known to geographers that its more general ramifications need no lengthy recapitulation. Of course, the definition of the area under consideration is often an essential beginning to a geographical study. Beyond this, the definition of relevant problems, the research procedures adopted, and the conclusions reached will vary in emphasis and probably in meaning, depending on the scale adopted. These matters are particularly important in population geography, because of an observed regularity in human behaviour. This is that migration to and from a place varies inversely with the distances involved. As a result, if population information is presented by large areal units, the volume of migration will appear to be less than if the areal mesh is fine. Thus the amount of migration observed depends on the scale of the areal units employed. The number of migrants to the whole of a city, for example, would be less than the sum of migrants to each of the wards or parishes into which the same city might be divided.

Migration and natural change are the two components of population change. Their relative importance will vary with the scale of the units under consideration. At the world scale, population changes result from the balance of births and deaths; at the continental level, and in most cases at the national level, natural change vastly outweighs migrational change in numerical importance. At the local scale (e.g. an English county) migration is much more important and it is usually dominant at the level of the street or block. Figure 9 shows the relative dominance of natural change and net migration between 1951 and 1961 for England and Wales at four different scales.

Cultural Differentiation

The scale issue appears in another familiar way. Unless a population study overlaps a cultural boundary, it is unlikely that a study of a small region will

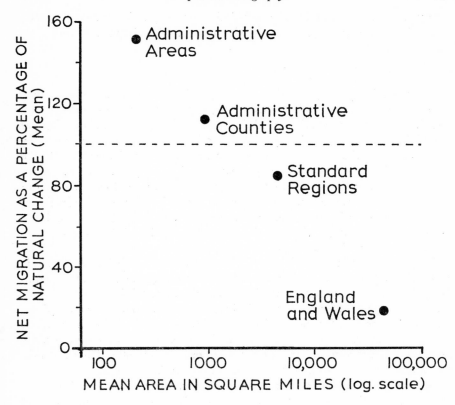

Fig. 9. Scales of division of England and Wales and net migration as a factor in population changes, 1951–61.

be concerned with fundamental differences in culture. It may, of course, deal with differences of age, sex, class, occupation, distribution, and density of population within a specific culture. By way of contrast, in a study of population changes in the lands around the Mediterranean Sea, the problems of spatial analysis will be dominated by the distinctions to be drawn between locational factors in thoroughly-different cultural regions.

Cultural differences are fundamental locational variables. How culture groups and the regions they occupy are to be defined is a matter for debate, but they cannot be ignored. Different ways of life are bound to affect basic characteristics of population. Population theories based on the record of the industrialized West are mostly inadequate for application to demographic systems in Africa and Asia. Attitudes to life and death, marriage and birth, family relationships and location are among the most discriminatory of cultural traits. As a result the population geographer's tool-kit will vary according to the cultural milieu in which his problem is located.

Population Data

We are all beset by statistics, the population geographer more so than others. It is a cultural trait of modern western man to collect and consume numbers, and more statistics have been collected and published about human populations than about anything else. The information is often in the finest detail, both in the features catalogued and the areal units employed. The plethora of numbers is both an advantage and a danger. The advantage is that data are available for most parts of the world so that research into fundamental questions is possible. Even in countries where the collection of statistics is in its infancy or where data are not available to the world at large, students of population will often be better provided with information than any other group of research workers.

The snare lies not so much with the huge deposit of statistical ore but in its refinement. The population geographer may get bogged down in the mining operation and produce only statistical descriptions containing lists of numbers in tabular or verbal form and descriptive maps, but with little or no explanation of the areal or locational characteristics that emerge. In this situation the development of theory or generalizations may be retarded or put aside, while inordinate amounts of time are spent on refining data. The propensity of the lay public to swallow large doses of population data without appreciable analysis or synthesis adds to the difficulty. Internal data problems are common, especially since information is usually collected and published for the administrative units which existed at the time of enumeration, and this implies difficulty in comparison over periods during which there have been revisions of local government boundaries.

Population Distribution

The distribution of population is a useful starting point for analysis. The production of accurate and detailed population maps involves considerable research and ingenuity (e.g. Burgdorfer, 1954; Zelinsky, 1966, pp. 8–9; Porter, 1966). Geographers are addicted to such maps not so much because of the distribution patterns themselves, but because of the associations that the maps stimulate in a well-stocked geographical mind. The development of expertise in recognizing these associations is a useful device for the geography teacher, provided the distribution maps are carefully selected and the teacher is prepared to demonstrate various ways of analysing the distributions.

World or continental distribution maps show heavy concentrations of population in East Asia, South Asia, and Europe, lesser groupings in North America and Southeast Asia, and more isolated clusters in Latin America, Africa, the Middle East, and elsewhere. Many areas of land are almost empty. Attempts to explain these fundamental distributions can be an exercise in futility—it is better to discuss the issues involved in accounting for them.

Overall explanations of world population distribution may appeal to some general theory such as that of "climatic determinism" (or a combination of climatic and terrain controls), of stages of economic development, or of political stability or instability as explanation of the presence or absence of population. But enough is known about the background to population distribution in different parts of the world for us to be sceptical of general solutions to the question of why people are concentrated in some areas and are sparsely settled elsewhere. Since culture, with all its variety, intervenes in the relationship between man and land, we can hardly expect that an accounting devoid of any cultural component will suffice. Further, since the effects of cultural differences on the population map are not instantaneous but are accumulated through time, an adequate study of the population distribution in any region must involve retrospection. Thus the broad facts of population distribution in China can only be dealt with by discussing, among other things, the slow build-up of numbers associated with the development and spread of a complex subsistence agricultural system, the emergence and perpetuation of a city-based administrative hierarchy, and the expansive effects of European contacts on populations along coasts and navigable waterways.

Population Change

A population distribution map is like a snapshot of a crowd. Individuals and groups are fixed in location at an instant of time. If we could add the time dimension to our maps, so that they also recorded births, deaths, and migrations, their surfaces would be as restless as the oceans. To many, the mutability of populations is their most interesting feature, not only in the short term with additions and subtractions within a social and technological system, but also in the long term, as the nature of society changes because of external influence or internal development. That Garden City and Dullsville each have 5000 inhabitants are facts of some interest. That one is increasing in population and the other declining is more relevant, because the types and rates of change are the summary result of many forces and pressures on the group and the individual, and a shorthand expression of the nature of geographical change itself.

From what has been said about the autonomy of population geography, it follows that analysis preferably should involve more than discursive accounts in which population changes are related informally to a wide variety of factors. More rigorous studies should deal quantitatively with the elements and connections within a population system. When these have been isolated their locational ramifications will be more meaningful. Thus while absolute populations and total population changes remain of much interest, especially through the regular publication of censuses, most of the research literature is concerned with one or other of the components of population change. These components are essentially pairs of alternatives: total change results from natural change and net migration; natural change from births and deaths; net migration from in-migration and out-migration. For comparative purposes these components acquire meaning when related to the population during the period under consideration; hence the importance of various indices such as birth rate, death rate, fertility rate, infant mortality rate, and migration ratios. These various descriptions of population change can be further analysed by comparisons between places that involve the consideration of (1) the internal structure of the population concerned (for example, numbers at different ages, proportions of women of childbearing age, and causes of death), and (2) the social and environmental system that impinges on the population (for example, public health measures, opportunities for migrants, and propaganda about the optimum number of children per family). Generally speaking, the relations between population changes and the internal structure of the populations concerned are easier to grasp and manipulate than correlations with external factors. However, since it is the mutual interrelations of the two over time that bring about change, neither can be discarded.

The Demographic Transformation

Population data for some western countries over the past two centuries reveal that the economic and social processes of modernization accompanied changes in fertility and mortality that led to large increases in population. In A.D. 1650 the total population of the European culture area was about 120 million, or only 22 per cent of the world's total; by 1920 the figure was about 700 million, or 40 per cent of the total. The population of England and Wales multiplied six-fold between 1750 and 1900, despite much emigration. Similar massive increases occurred in other countries, including immigrant areas like the United States and emigrant countries such as Sweden (Broek and Webb, 1968, pp. 428–57). The changes in mortality and fertility during this period of transformation are well known, and various theories have been proposed to generalize their relationship to changing production of resources (Wrigley, 1967).

In the "traditional society", before the yeast of modernization had begun its work, high fluctuating birth and death rates restrained the expansion of population; periods of slow growth in settled times were followed by massive losses from warfare, disease, or famine. In western Europe, from the mid-eighteenth century onwards, mortality fell slowly as food production increased and was better distributed, and public health measures found gradual acceptance. Births continued at a high rate and continued population growth was established. Later, fertility began to decline; income from child labour decreased and as more children survived, the need for having many disappeared. New contraceptives helped to reduce family sizes for those who saw many children as a positive disadvantage in a society where an expensive education during a long non-working youth became essential and where women were no longer willing to accept the status of mere child-bearing drudges. Eventually birth rates fell to just above death rates and population growth declined.

This is the historical record, but can the experience of European countries be turned into a theory applicable to non-European areas? Will all societies experience massive population growth as they undergo modernization? The evidence is unclear. Death control, introduced so successfully into many parts of Africa, Asia, and Latin America in the mid-twentieth century, has opened an awesome gap between fertility and mortality that is bringing increases twice or three times as high as was the case in Europe. The key to the applicability of the European-based hypothesis now lies with birth control and the reduction of fertility. Since disease control was successfully injected into the life of many peoples without the paraphernalia of modernization, is a similar breakthrough likely for fertility? There is no answer to this question yet, but on it depends the nature of life for most of mankind as we approach the twenty-first century.

Evidence in the form of demographic data from such sources as the United Nations suggests that birth control can be introduced into partly modernized societies such as Taiwan and Singapore, but that there is much greater variation in its acceptance than with death control. The desire to avoid death is nearly universal, but responses to the possibility of family limitation vary widely according to deeply held beliefs embedded in the different cultures.

The "Population Problem"

The relationship of current or projected population increases to the production of food or other resources constitutes the "population problem". On this issue there is much confusion in the public mind, for expert opinion on the subject is highly varied. Some scholars, echoing the ideas of T. R. Malthus, argue that if increases continue at present rates, the earth will be unable to supply adequate food and fibre and other necessities, and mass starvation and strife will result. Optimists point out that if known techniques of organization, production, and distribution could be applied over much of the earth, then a population many times larger than the present one could be supported (Population Reference Bureau, 1968).

For the geographer little can be accomplished at these rarified levels of argument. The smallest effort at analysis shows that the relationship of people to resources is one of great variety. The geographical approach offers advantages. As the population and resources question is considered in specific areas it becomes clear that a single world view has little value. The varied cultural milieux dispel notions of "optimum population". This term, with "overpopulation", "underpopulation", and "carrying capacity" is seen to be useful in only a relative sense, relative to particular peoples and their habitats. Zelinsky (1966) and Broek and Webb (1968) showed, by way of case studies, that the relationship between population change and material welfare is highly varied. In a thoroughly modernized country such as Sweden resources imply items like output and investment, educational systems, and welfare arrangements. In India, by contrast, the crucial matters are alleviation of poverty, the production of food, and its distribution within a fundamentally village society. The reasons behind population changes and the changes themselves are quite different.

It might be argued that the national scale is too gross to reveal any conclusions for population geography. Indeed, the focus of most geographical work has been local and regional rather than national or continental. However, the national scale becomes worthwhile when one realizes that in a modernized society the events that make up the population system occur within the areal context of the national state. National economy, national plans, national health schemes, national population redistribution policies all imply that investigation of population developments can be carried on effectively at the national level. National borders are becoming regional boundaries between population systems.

Local Populations

The growing importance of national policies does not, however, result in the disappearance of local and provincial differences within individual countries. On the contrary there is much evidence to show that modernization brings in its train new and sharp differences in local population numbers and characteristics. Thus in England and Wales the Industrial Age was accompanied by major changes in the distribution of population. By the end of the nineteenth century, typical population maps (e.g. Lawton, 1968) reveal two demographic types: (1) industrial-mining concentrations with massive populations which were relatively youthful, densely settled, experiencing high mortality and fertility rates, and added to by steady migrational streams; and (2) farm, village, and small-town groupings with generally older age structures, lower fertility and mortality rates, and growing slowly or declining because of many decades of out-migrations to the industrial areas. The percentages of population change of the two types over short time periods were not very different, since the margins between births and deaths were roughly similar. Over the long term, however, the cumulative effects of migrations from rural and small-town districts to industrial regions built up large numbers of people in the latter areas. A broad explanation of these patterns is the development of the industrial-mining economy and the relative decline of farming and small-town activities as means of livelihood. The characteristics of these demographic types and the impact of the division on national life formed the subject of many contemporary statistical treatises (e.g. Welton, 1911).

The locational features of population changes in England and Wales in the late twentieth century are quite different. Under the welfare state, fertility and mortality are much reduced, and the rise of personal affluence has allowed migration to become part of everyday living. Mobility is characteristic of the young going to a new job or marriage, of the established adult moving to a better suburb or perhaps a new town, of the aged going to a retirement town or a place in the country. Migrations are thus age-differentiated as to destination. Moreover, the young tend to migrate toward the central part of the country and the old toward the coastal periphery, resulting in the sharp distinctions in age composition epitomized by the tens of thousands of young adults and their children in new towns and suburbs and the tens of thousands of retired folk and widows in places like Southport and Bournemouth. The range of natural change is wider than it used to be; places with high fertility have few deaths, and places with many deaths tend to have few births. The system is maintained by continuing migration which provides new inhabitants for city centre, suburb, and coastal town, as individuals pass through the locational stages of their lives.

It is perhaps instructive to realize that the changes which geography itself has undergone in the past two generations are analogous to a fundamental change in the locational nature of the general population. In the age when the primary determinants of population distribution and change were the sites of resource production and their varied social and environmental conditions, it was appropriate that geographers' explanations should involve local conditions. Present-day geography's preoccupation with networks, flows, and linkages within systems fits well with population distributions and changes that are increasingly dominated by situation or relative location within a life–death system in which mobility plays the dominant rôle.

REFERENCES AND FURTHER READING

The growing body of methodological and substantive literature concerned with population geography includes:

BEAUJÈU GARNIER, J. (1956 and 1958) *Géographie de la Population*, Librarie de Médicis, Paris, 2 vols.

BEAUJÈU GARNIER, J. (1966) *Geography of Population*, Longmans, London.

CLARKE, J. I. (1966) *Population Geography*, Pergamon, Oxford.

GEORGE, P. *Questions de la Géographie de la Population*, P.U. de France, Paris.

Political and Economic Planning (1955) *World Population and Resources*, P.E.P., London.

SAUVY, A. (1952 and 1954) *Théorie Générale de la Population*, P.U. de France, Paris, 2 vols.

VIDAL DE LA BLACHE, P. (1922) *Principes de Géographie Humaine*, Colin, Paris, reprinted in English in 1956 by Constable, London.

ZELINSKY, W. (1962) A bibliographic guide to population geography, *University of Chicago Geography Research Paper*, **80**.

ZELINSKY, W. (1966) *A Prologue to Population Geography*, Prentice-Hall, Englewood Cliffs.

The nature and prospects of population geography are considered in:

POKSHISHEVSKIY, V. V. (1962) Geography of population and its tasks, *Soviet Geography*, **3**, 3–13, November.

TREWARTHA, G. T. (1953) A case for population geography, *Annals of the Association of American Geographers*, **43**, 71–97.

RECENT DEVELOPMENTS IN POPULATION GEOGRAPHY

HOOSON, D. J. M. (1960) The distribution of population as the essential geographical expression, *Canadian Geographer*, **17**, 10–20.

WRIGLEY, E. A. (1965) Geography and population, in CHORLEY, R. J. and HAGGETT, P. (Eds.) *Frontiers in Geographical Teaching*, Methuen, London, pp. 62–80.

WRIGLEY, E. A. (1967) Demographic models and reality, in CHORLEY, R. J. and HAGGETT, P. (Eds.) *Models in Geography*, Methuen, London, pp. 189–215.

SOME ISSUES IN POPULATION GEOGRAPHY

Valuable sources of population data include:

United Nations (1949 to date) *Demographic Yearbook*, U.N.O., New York.

United Nations (1949 to date) *Statistical Yearbook*, U.N.O., New York.

WITTHAUER, K. (1958) Die Bevölkerung der Erde: Verteilung und Dynamik, *Petermans Geographische Mitteilungen, Supplement* 265.

Solutions to the problem of mapping population data are to be found in:

BURGDORFER, F. (Ed.) (1954) *World Atlas of Population*, Falk, Hamburg.

HOWE, G. M. (1963) *National Atlas of Disease and Mortality*, Nelson, Edinburgh.

HUNT, A. J. (Ed.) (1968) Population maps of the British Isles, *Transactions of the Institute of British Geographers*, **43**.

PORTER, P. W. (1966) East Africa—population distribution, *Annals of the Association of American Geographers*, **56**, map supplement no. 6.

STEWART, J. Q. and WARNTZ, W. (1958) The physics of population distribution, *Journal of Regional Science*, **1**, 71–97.

A representative study of Swedish theoretical contributions to the study of population change is:

OLSSON, G. (1965) Distance and human interaction, a migration study, *Geografiska Annaler*, series B, 3–43.

Population change is also considered in:

SORRE, M. (1955) *Les Migrations des Peuples: essai sur la mobilité géographique*, Flammarion, Paris.

Studies related to the "demographic transformation" include:

BROEK, J. O. M. and WEBB, J. W. (1968) *A Geography of Mankind*, McGraw-Hill, New York, pp. 420–96.

CARR-SAUNDERS, A. M. (1936) *World Population: past growth and present trends*, Oxford U.P., Oxford, reprinted 1966.

CIPOLLA, C. (1962) *The Economic History of World Population*, Penguin, Harmondsworth.

United Nations (1956) Population growth and the standard of living in under-developed countries, *U.N. Population Studies*, **20**.

A useful contribution to the "population problem" is:

Population Reference Bureau (1968) The population–food dilemma, *Population Bulletin*, **24**, 81–99.

Local populations, in the British context, are studied in:

LAWTON, R. (1968) Population changes in England and Wales in the later nineteenth century: an analysis by registration districts, *Transactions of the Institute of British Geographers*, **44**, 55–74.

VINCE, S. W. E. (1952) Reflections on the structure and distribution of rural population in England and Wales, 1921–1931, *Transactions of the Institute of British Geographers*, **18**, 53–76.

WEBB, J. W. (1963) The natural and migrational components of population changes in England and Wales, 1921–1931, *Economic Geography*, **39**, 130–48.

WELTON, T. R. (1911) *England's Recent Progress*, Chapman & Hall, London.

WRIGLEY, E. A. (1961) *Industrial Growth and Population Change*, Cambridge U.P., Cambridge.

CHAPTER 10

EARLY MAN AND ENVIRONMENT

C. VITA-FINZI

AN ADVERTISEMENT for a recent book stated that it reflects a "highly significant trend in modern geographical studies" in that it stresses the influence of
the environment on the history and culture of the peoples in question. This
trend began over 2000 years ago and the interaction between man and his
environment has ever since formed an integral part of geographical education.
Yet the tendency has been to refer only to case-studies drawn from historical
times, prehistory being regarded as the preserve of archaeologists and Quaternary geologists. The time has come to review this demarcation, whose arbitrary character is increasingly made obvious by the growing contribution
made to the study of early man by geographers and geographical techniques.

At first glance the study of pre-literate times might even seem to offer
certain advantages, notably a directness in the links between man and nature
which cultural "accretions" tend increasingly to confuse, and a body of data
small enough to be mastered; but the former of these often turns out to be
illusory, and the latter, when applicable, merely restricts the scope of the
investigation. It is probably truer to say that, for the geographer, the main
virtue in extending the bounds of human and historical geography into
prehistory lies in the opportunity it provides for fusing together socio-
economic concepts with those of physical geography. This can also be claimed
for studies of the recent past; but here the environment tends to be seen as
a backdrop against which technology and society evolve and which, barring
minor fluctuations of climate and the localized problem of soil erosion, is
essentially static (van Bath, 1963). In the long-term view, however, all
elements of the physical environment are seen to have changed.

But there is no need to justify what is, after all, a manifestation of man's
curiosity about his origins. Attention will be directed here to topics which are
of interest either because they are being actively pursued or because they
have special pedagogical virtues.

MAN AND NATURE

Philosophers have been struggling with the problem of "free will" long and fruitlessly enough for the geographer to recognize at last a field in which he can afford not to tread. But trod he has, as the prolonged debate over the validity of environmental determinism bears witness (Tatham, 1951). The debate now appears to have subsided, and if attempts to provide comprehensive statements of man's relationship to his environment—such as Toynbee's (1934) "Challenge-and-Response"—are any longer mentioned, it is almost with embarrassment. Determinism as crude as that preached by Semple (e.g. 1947) and Huntington (e.g. 1922) is outmoded, and its more palatable alternatives, "possibilism" and "probabilism", have served for little more than terminological wrangles. It is illogical to qualify determinism, but no less nonsensical to describe man–land relationships in simple cause-and-effect terms.

Yet certain attitudes and beliefs continue to be peddled implicitly. Perhaps in reaction to the Dust Bowl and to the widespread phenomenon of soil erosion, there developed in the 1930's a sense of guilt that survives in loaded terms such as "accelerated" erosion, and that brands modern man, like the goat, as invariably an agent of destruction (Thomas, 1956, *passim*). Martin (1967, pp. 109–10) quoted an example of the " 'rose-coloured' view of prehistoric man in Africa", which holds that during his occupation of the continent and prior to the introduction of modern weapons, "animal life was, in a large sense, in a virtual state of equilibrium"; his own view is that the extinction of many large vertebrates that took place at the close of the Ice Age was due to "overkill" by prehistoric man (Martin, 1967). And it is not only in the sphere of explanation that the newcomer to the subject will be confronted with doctrinaire statements. Every technique breeds its dogma, and its practitioners may come to view any criticism as an inroad on their prestige. Hence pollen analysis, radiocarbon dating, oxygen isotope studies, these and other tools for the investigation of past environments stand to benefit from periodic re-evaluation by uncommitted outsiders (e.g. Tauber, 1967; Raikes, 1967).

In the formulation of the problems themselves the rôle of preconceptions may be equally important. Take animal "domestication", a theme which is profoundly interesting to zoologists and archaeologists: what if it should emerge as a fiction, in that symbiotic associations between man and the local fauna could have prevailed well before its supposed introduction (Higgs and Jarman, 1969)? The issues have been complicated by the habitual isolation of man from the rest of the organic world. We see this nineteenth-century habit persisting, in the teeth of Darwinism, whether the theme be man's adaptation to his environment (e.g. Childe, 1941) or the evolution of man's consciousness (Teilhard de Chardin, 1959). It is, in effect, the standpoint of the "culture history" school of archaeology (Flannery, 1967); the opposing "process" school thinks it futile to speak of an environment "outside" culture, for man participates in numerous systems which comprise both cultural and non-cultural phenomena. (As Flannery pointed out, the process approach continues the trend towards determinism begun by the culture historians, since "decisions" are moved even further away from the individual's sphere of influence: the systems, once set in motion, are self-regulating.) At what point in pre- or proto-history does an anthropocentric view first become feasible? And is it ever justified?

It is worth noting, in passing, Flannery's prediction (1967) that general systems theory, game theory, and locational analysis would be applied successfully to American archaeology within the next decade. One year later a book exploiting these and analogous approaches (Clarke, 1968) is already causing concern among more conservative British archaeologists, much as Chorley and Haggett (1965) recently did among geographers; yet, ironically enough, it is the culture history school which is best served by it, since Clarke's definition of archaeology is "the study of ancient artifacts".

Whatever the standpoint, certain items of information relating to the environment are needed, and by providing these the geographer can advance beyond the rôle of devil's advocate in the evaluation of other sources of information. Three possible approaches will be discussed below.

DATING THE LANDSCAPE

The reconstruction of the landscape at different periods in the past, or palaeogeography, is commonplace in geology. A combination of geological and archaeological techniques allows this to be done for prehistoric times. The resulting information is of immediate value in interpreting the topographical context of sites. And, as archaeologists become increasingly interested in the economy of early man, so knowledge relating to the availability of resources in the past becomes the more desirable. Hence the relevance of studies which date the deposition of soils or the incision of valley floors or which chronicle the development of a delta.

Archaeological dating is not new. Boucher de Perthes' discovery of flint implements associated with fossil animal remains in the Somme valley took place in 1838 (Daniel, 1950); by the 1860's artifacts were already being used by geologists to work out the chronology of valley-cutting by streams (Chorley *et al.*, 1964, pp. 447–8). The danger of circular argument (through the dating of deposits by artifacts dated geologically elsewhere) led Zeuner (1959) to attempt a stratigraphic study of the Pleistocene period divorced from archaeology; while Oakley (1964, p. 9), who recognized the value of archaeological dating, limited it to relative (as opposed to absolute) dating. But it is sometimes possible, even when dealing with the pre-pottery period, to obtain a closely dated chronology for the geological development of an area, provided distinctive artifact assemblages are found both in association with the successive landforms and within stratified cave sediments which have been dated by means of carbon-14 or other radiometric methods. In some instances individual artifacts may suffice, although archaeologists, despite elaborate statistical studies of large samples, are increasingly reluctant to employ the term "typical" or "diagnostic". The ideal is an industry which evolved rapidly, spread quickly, and was soon supplanted.

Thus the first problem is that of finding suitable "zone fossils". The second is stratigraphic; the artifact must be in place, within a formation demarcated on other, geological, grounds. Even when there is no suspicion of intrusion by earthworms or rodents, however, the artifact only provides us with a

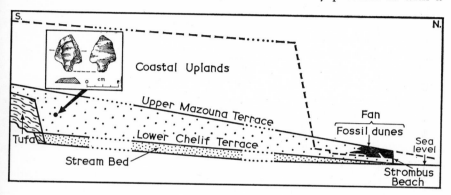

FIG. 10. Longitudinal profile of a coastal wadi in Algeria, showing two alluvial fills and their relationship to present sea-level and the raised beach of the last Inter-glacial which includes the fossil *Strombus*. The inset shows the Aterian point found within the Mazouna (older) fill.

minimum date; hence the desirability of combining archaeological with radiocarbon dating. These two techniques are illustrated by the use of a single Aterian point in dating a wadi deposit in Algeria (Fig. 10) and of both Upper Palaeolithic artifacts and carbon-14 analysis in dating the plain of Wadi Hasa, Jordan (Vita-Finzi, 1966 and 1967). The technique finds greatest scope in the study of alluvial deposits, which both harbour and yield artifacts willingly; but it has been applied successfully to spring-laid tufas (McBurney and Hey, 1955, pp. 142–59), to folded formations (Stekelis *et al.*, 1960) and to many other situations.

By these means the extent and character of geological change in prehistory can be established, and fluctuations in "available land" evaluated (e.g. Higgs and Vita-Finzi, 1966). But this is unlikely to suffice, especially where pre-agricultural societies are involved, for the potential of any exploitable area depended very indirectly on the geological setting. The record has to be interpreted in a manner which yields information regarding other environmental elements.

ENVIRONMENTS, LOCAL AND REGIONAL

What Butzer (1964) has termed Pleistocene geography is an attempt to fuse together the environmental evidence of a wide variety of indicators, geological, botanical, zoological, and so forth. Unfortunately the common language of all these disciplines has traditionally been climate, even where the climatic interpretation of each of the indicators is highly dubious. Another problem which stems from a climatic approach is the temptation it provides to read into the defective local record whatever general chronology the investigator believes in. Thus adherents to the traditional four-fold glacial scheme continue to find traces of four glaciations wherever they look, or are forced to conjure up an explanation for the absence of any one of them.

The archaeologist is now primarily interested in the environment of his site and its immediate vicinity, and no longer in large-scale climatic fluctuations which might serve to explain tribal migrations or the fate of civilizations. What is more, this is to be expressed not in meteorological language, but as statements regarding the depth of the water table, the character of the soil, the régime of the local streams, etc. Fortunately, morphological, sedimentological, palynological, and other clues to the past reflect local factors such as drainage, aspect, and relief more than they do zonal climate.

It is, of course, important to define "local" when delimiting that portion of the environment which is relevant to an archaeological site. Contemporary studies of territorial limits among different species (including hunting man) give some indication of the "catchment" of certain kinds of site. Similarly if there is evidence of seasonal or other rhythmic environmental oscillations during the period of human occupation, the archaeological and environmental record may benefit from being interpreted in the light of transhumance and analogous responses to such alternations. Sites which were regarded as the products of different "cultures" may thereby emerge as specialized camp-sites fulfilling different functions (Higgs *et al.*, 1967).

RATES OF EROSION AND DEPOSITION

So far emphasis has been laid on the interpretation of the human record, and on the contribution which geographers and their colleagues can make to it. But the earth sciences also stand to gain from interdisciplinary collaboration.

Thus archaeologically dated alluvial stratigraphies can yield useful information on the rate at which erosion and deposition proceeded in past times. These figures can then be compared with present-day estimates of rates based on direct measurements, data on reservoir sedimentation, etc. The result may be illuminating where, as in Wyoming (Leopold and Miller, 1954), ancient erosion rates are found to differ little from those of the present day. The lesson for the conservationist is then clear, if not very helpful: the human factor in this particular instance sheds some of the significance habitually ascribed to it.

It may ultimately prove possible to link erosional and depositional episodes with phases of ecological change directly, that is to say without climatic interpretation, by means of Erhart's scheme of *biostasie and rhexistasie* (Erhart, 1956). Briefly, these terms denote, respectively, a state of biological stability and one of instability; the latter may be precipitated by man-made changes in the vegetation, climatic fluctuations, and earth movements.

Clearly there is ample room for further collaboration between the geographer, the prehistorian, and all those concerned with the social and economic history of man. The rôle of the geographer—especially now that he aspires to be more than a professional dilettante—will continue to gain in importance. The three issues discussed above may have succeeded in showing that this increased participation provides the teacher with ample material for discussion and, what is more, with a unifying theme for fieldwork: man's changing response to a changing environment, and his contribution to those changes.

REFERENCES AND FURTHER READING

A comprehensive if somewhat gloomy general introduction and reference text is:

THOMAS, W. L. (Ed.) (1956) *Man's Role in Changing the Face of the Earth*, U. of Chicago P. Chicago.

For some discussion of the economic significance of environmental change, see:

VAN BATH, B. H. S. (1963) *The Agrarian History of Western Europe, A.D. 500–1850*, Arnold London.

MAN AND NATURE

CHILDE, G. (1941) *Man Makes Himself*, Watts, London.
CHORLEY, R. J. and HAGGETT, P. (Eds.) (1965) *Frontiers in Geographical Teaching*, Methuen, London.
CLARKE, D. L. (1968) *Analytical Archaeology*, Methuen, London.
FLANNERY, K. V. (1967) Review of WILLEY, G. R. (1967) *An Introduction to American Archaeology*, Prentice-Hall, Englewood Cliffs, *Scientific American*, **217**, 119–22.
HIGGS, E. S. and JARMAN, M. (1969) The origins of agriculture: a reconsideration, *Antiquity*, **43**, 31–41.
HUNTINGTON, E. (1922) *Civilization and Climate*, 2nd edn., Milford, London.
MARTIN, P. S. (1967) Prehistoric overkill, in MARTIN, P. S. and WRIGHT, H. E., JR. (Eds.) *Pleistocene Extinctions*, Yale U.P., London, pp. 75–120.
RAIKES, R. L. (1967) *Water, Weather and Prehistory*, Baker, London.

SEMPLE, E. C. (1947) *Influences of Geographic Environment*, Constable, London.
TATHAM, G. (1951) Environmentalism and Possibilism, in TAYLOR, G. (Ed.) *Geography in the Twentieth Century*, Methuen, London, pp. 128–62.
TAUBER, H. (1967) Differential pollen dispersion and filtration, in CUSHING, E. J. and WRIGHT, H. E., JR. (Eds.) *Quaternary Paleoecology*, Yale U.P., London, pp. 131–42.
TEILHARD DE CHARDIN, P. (1959) *The Phenomenon of Man*, Collins, London.
TOYNBEE, A. J. (1934) *A Study of History*, **1**, Oxford U.P., London.

DATING THE LANDSCAPE

CHORLEY, R. J., DUNN, A. J. and BECKINSALE, R. P. (1964) *The History of the Study of Landforms*, **1**, Methuen, London.
DANIEL, G. E. (1950) *A Hundred Years of Archaeology*, Duckworth, London.
HIGGS, E. S. and VITA-FINZI, C. (1966) The climate, environment and industries of Stone Age Greece: Part II, *Proceedings Prehistoric Society*, **32**, 1–29.
McBURNEY, C. B. M. and HEY, R. W. (1955) *Prehistory and Pleistocene Geology in Cyrenaican Libya*, Cambridge U.P., Cambridge.
OAKLEY, K. P. (1964) *Frameworks for Dating Fossil Man*, Weidenfeld & Nicolson, London.
STEKELIS, M., PICARD, L., SCHULMAN, N. and HAAS, G. (1960), Villafranchian deposits near Ubeidiya in the central Jordan valley (preliminary report), *Bulletin Research Council of Israel*, **9G**, 175–83.
VITA-FINZI, C. (1966) The Hasa Formation, *Man*, **1**, 386–90.
VITA-FINZI, C. (1967) Late Quaternary alluvial chronology of northern Algeria, *Man*, **2**, 205–15.
ZEUNER, F. E. (1959) *The Pleistocene Period*, Hutchinson, London.

Dating and other techniques are also described in:

BROTHWELL, D. R. and HIGGS, E. S. (Eds.) (forthcoming) *Science in Archaeology*, 2nd edn., Thames & Hudson, London.
ZEUNER, F. E. (1958) *Dating the Past*, Methuen, London.

ENVIRONMENTS, LOCAL AND REGIONAL

BUTZER, K. W. (1964) *Environment and Archeology*. Methuen, London.
HIGGS, E. S., VITA-FINZI, C., HARRIS, D. R. and FAGG, A. E. (1967), The climate, environment and industries of Stone Age Greece: Part III, *Proceedings Prehistoric Society*, **33**, 1–29.

Regional examples:

BRYAN, K. (1954) The geology of Chaco Canyon, New Mexico, *Smithsonian Miscellaneous Collection*, **122**, no. 7.
BUTZER, K. W. and HANSEN, C. K. (1968) *Desert and River in Nubia*, Wisconsin, U.P. Madison.
SMALLEY, I. J. (1968) The loess deposits and Neolithic culture of northern China, *Man*, **3**, 224–41.
WRIGHT, H. E., JR. and FREY, D. G. (Eds.) (1965) *The Quaternary of the United States*, Princeton U.P., Princeton.

For treatment of specific elements of the environment see:

CORNWALL, I. W. (1958) *Soils for the Archaeologist*, Phoenix House, London.
KRUMBEIN, W. C. and SLOSS, L. L. (1963) *Stratigraphy and Sedimentation*, Freeman, London.
LEOPOLD, L. B., WOLMAN, M. G. and MILLER, J. P. (1964) *Fluvial Processes in Geomorphology*, Freeman, London.
MANTEN, A. A. (1967) Palynology and environmental geology, *Palaeogeography, Palaeoclimatology, Palaeoecology*, **3**, 7–15.
NAIRN, A. E. M. (Ed.) (1961) *Descriptive Palaeoclimatology*, Interscience, London.

SAWYER, J. S. (Ed.) (1966) *World Climate from 8000 to 0 B.C.*, Royal Meteorological Society, London.
UNESCO (1963) *Changes of Climate*, UNESCO, Paris.

See also:

CALDWELL, J. R., (Ed.) (1966) *New Roads to Yesterday*, Thames & Hudson, London.

RATES OF DEPOSITION AND EROSION

ERHART, H. (1956) *La Genèse des Sols en tant que Phénomène Géologique*, Masson, Paris.
LEOPOLD, L. B. and MILLER, J. P. (1954) A postglacial chronology for some alluvial valleys in Wyoming, *U.S. Geological Survey Water-Supply Paper*, **1261**.

CHAPTER 11

PROGRESS IN HISTORICAL GEOGRAPHY

H. C. PRINCE

THE focus of interest in historical geography has shifted from time to time across a broad range of inquiries, now clarifying one aspect, now another. Progress has been achieved not so much by following a single path from one level of generalization to another but by approaching problems from different angles, bringing to light new relationships between known and hitherto unknown facts and ideas. Works published under the title of historical geography show how diverse these approaches are; they include gazetteers of the ancient world, accounts of the influence of geography upon history, reconstructions of past geographies, studies in sequent occupance, chronicles of geographical change through time, retrogressive narratives, investigations of relict features, critical appreciations of the perceptions of the past, and essays in theory (Darby, 1953; Smith, 1965). Some of these inquiries have been concluded or abandoned, others are just beginning. Most of the scenes of major events in world history have been gazetteered and inscribed on maps. The part played by geographical influences on the course of history is no longer treated deterministically. In the study of sequent occupance new objectives are being sought, and frontiers of settlement are being examined by new methods. The investigation of past geographies, of geographical change and of the past in the present are still major activities, while fresh approaches to historical geography through the perception of the past and through models are now being explored.

PAST GEOGRAPHIES

European historians have led the way in describing geographies of the past in order to set historical actions in appropriate geographical contexts. Braudel gave the title *géohistoire* to his broad introductory survey (Braudel, 1949). At each step back in time history becomes more geographical until, at the furthest reach of human history, it is all geography. There lies the primitive landscape, the *Urlandschaft* (Gradmann, 1898). From studies of prehistory, mainly from the work of archaeologists, geographers have acquired two ideas.

The first is that the "datum line from which change is measured is the natural condition of the landscape" (Sauer, 1925, p. 37); the second is that the personality of a country is moulded at "formative periods in a distant past" (Sauer, 1941a, p. 354). Neither view can be accepted without reservation until all potential formative periods, recent as well as remote, have been studied, and until the obscure dawn of human history is more fully illuminated. In the Old World primitive landscapes are so remote in time that only imperfect pictures can be drawn, but in the United States the appearance of the public domain on the eve of white settlement is recorded in the plats and notes of the Federal Land Survey. Even where documents are available, however, there are few large areas for which it is possible to reconstruct a primitive landscape at a single instant in time.

Those who share the views expressed by Hettner and Hartshorne (Hartshorne, 1959) that geography is concerned with the functional interrelationships of phenomena in space are committed to a study of areas at particular moments in time, generally at the present. Historical geography is assigned the task of restoring the "historical present", that is, the present as it existed at some moment in the past (Mackinder, 1930, p. 310). The causal interdependence of spatial arrangements is investigated for cross-sections in the past as it is for the present, except that evidence for the past is drawn from historical records. Instantaneous pictures may be reconstructed from old photographs, from maps, from census returns, and from inventories such as the Domesday Book (Darby, 1952). A true cross-section, however, may be unhistorical in being severed from the chains of events necessary to its understanding.

GEOGRAPHICAL CHANGE

Whatever is of interest in a momentary situation "is to be understood only in terms of the processes at work to produce it" (Clark, 1954, p. 71). When the music stops in a game of musical chairs the players find themselves in unexpected positions. To understand how they reached those positions it is necessary to know something about the speed and direction of their movements, about the rhythm of the music, and about how long the music has been playing. To find out how things came to be where they are, or were, at a particular time, it is necessary to study the work of agents generating change, development, movement; in a word, to follow a genetic approach.

A series of static pictures arranged chronologically will reflect, albeit inadequately, the extent of changes taking place, but the processes producing those changes will be unaccounted for or, at best, only implied by the temporal sequence. In attempting to read more into the passage of time, development may be assumed to proceed by stages. This is the approach taken by

Whittlesey, comparing sequent occupance to vegetation succession or to the cycle of erosion, in which each stage of occupance "carries within itself the seed of its own transformation" (Whittlesey, 1929, p. 162) and contributes something to its successor. Sequent occupance studies have employed many different criteria for dividing time, but have failed to provide coherent accounts of processes. Unlike plant communities or landforms, human landscapes are not evolved in isolation, they are shaped by complex processes of cultural transference, intermixture, absorption, and replacement.

In the frontier Turner found a theme drawing together the actions and interactions of many factors. In 1893 he announced that "the existence of an area of free land, its continuous recession and the advance of American settlements westward, explain American development" (Edwards and Mood, 1938, p. 186). A serious flaw in his argument is the interpretation of frontier history as an evolutionary progression from primitive to advanced economies in which hunting is succeeded by sedentary cultivation and ultimately by the building of cities. Neither in the United States nor in other parts of the world is such a regular succession observed.

Some historians and archaeologists solve the problem of writing explanatory narrative by tracing the development of individual features separately through time. This method was employed in Darby's account of the changing English landscape which treated themes chronologically one by one: the wood was cleared, the marsh was drained, the heath was reclaimed, the garden was landscaped, towns and industries were expanded (Darby, 1951). A long line of investigation stemming from *Man and Nature* (Marsh, 1864) examines man's work as an agent in geological and climatic change and, above all, in changing the flora and fauna of the earth (e.g. Thomas, 1956). In Europe historians and geographers have recounted the extension of cultivation by *défrichement*, the advance and retreat of viticulture, the effort of embanking and draining, monastic colonization, and the foundation of bastides. The work of comparing the development of agrarian structures in different countries has been stimulated recently by the appearance of several international publications, notably *Morphogenesis of the Agrarian Cultural Landscape* (Helmfrid, 1961). The term "morphogenesis" expresses succinctly, but somewhat inelegantly, the nature of these inquiries into the changing forms of landscape features.

Methods have yet to be devised to integrate the histories of isolated elements into the spatial entities to which they belong, to reconstitute spatial connections between separate histories. Much may be learned about the mechanism of changing patterns by studying the changes of several different elements together, as in changing crop and livestock combinations (Weaver, 1954). Population changes may be studied by analysing changes in the contributions made by rates of migration and by rates of natural change

(Webb, 1963). What are especially difficult to represent are differing rates of change for several elements through time (Clark, 1962).

THE PAST IN THE PRESENT

The present landscape provides many keys to the past. Field observations raise and may resolve questions upon which documents are silent. The extent of early modifications to the land surface may be gauged from changes that have occurred in recent times, or may be tested experimentally. The Kon Tiki expedition has proved the feasibility of crossing the Pacific on a balsa raft and the efficacy of flint axes for clearing woodland has been demonstrated by felling live trees with replicas of neolithic implements.

The reconstruction of the past from the present, by what Maitland called "the retrogressive method" (Maitland, 1897, p. 5) proceeds from the better known to the less well known, from the visible real world to the obscurity of the Dark Ages. Taking care not to overlook intervening changes it follows "the trail backwards, one careful step at a time, examining irregularities and variations as they come, avoiding the all too common error of trying to leap at a bound from the eighteenth century to the Neolithic age" (Bloch, 1966, p. xxx). Recent studies of the changing character of field systems in Kent have been conducted by deriving earlier forms from the more certain evidence of later forms, by projecting trends from the recent to the remote past (Gulley, 1960; Baker, 1966 and 1968).

All features in the present landscape are relict features, survivals from some past period. Constant reference to past events is necessary to understand how they came to occupy their present positions, but not all past events are equally important. The depth to which the past is fathomed depends on the level of causation being sought. Ogilvie pleaded that a retrospective study "be applied simultaneously to the entire human imprint within a selected area so that the elements of it may be classified tentatively as belonging to successive stages in time" (Ogilvie, 1952, p. 7). For a small area in Sussex a map dating the man-made features in the landscape has been compiled (Yates, 1960). Watson treated relict features not only in the present scene but as objects produced by continuing processes of change, considering them "indicators of the ever-moving frontiers of the past which make up the ecology of the present" (Watson, 1959).

Many studies treat the whole of the present as a heritage from the past, but Sauer took a narrower view of cultural relicts as "surviving institutions that record formerly dominant but now old-fashioned conditions" (Sauer, 1941b, p. 15). Such relicts include traditional forms of industrial organization, antique farming practices and field systems as well as visible remains both extensive and fragmentary. They may be chance survivals embedded among other puzzling vestiges of vanished epochs: the strange marks of ridge and furrow, strip lynchets, moated farmsteads, pits and ponds, parks and curious garden ornaments, and deserted village sites, to list only a few English examples.

There is rarely a moment when a landscape feature finally disappears. The process of transformation usually preserves the imprint of an earlier form: Elizabethan market places, medieval burgage plots, and Roman walls leave their marks on the ground plans of modern towns. In the United States the rectilinear grid drawn by the original land survey lingers in a hundred forms, as perhaps the most important characteristic of the lands west of the Appalachians (e.g. Thrower, 1966).

Viewed in the framework of the present-day economic and social geography of an area relict features are those which cause functional friction. They are obsolete and they are resistant to change. A state of industrial inertia is said to exist where innovation is adjusted to conform to a relict pattern, new activities gravitating towards established centres, existing structures affecting the value of their surroundings. Sheffield continues to manufacture steel long after the initial locational advantages have ceased to operate, London mews provide flats and studios long after horses have departed, frontages overlooking London parks attract high land values, while Chicago's cemeteries and stockyards depress neighbouring land values.

Relict features may also be preserved as monuments or outdoor museums of past cultures such as Stonehenge and the Acropolis, or places of period charm such as Bath and New Orleans. The preservation of green belts around cities, of National Parks and areas of special landscape or historic value at once fossilizes large areas and deflects new development to unprotected localities.

PERCEPTION OF THE PAST

Once a historical geographer asks the question "why?" the full answer can no longer be obtained by consulting the record of man's works in the landscape because human motives, attitudes, preferences, and prejudices must also be examined. The development of ideas, the growing awareness of new realms of experience, the reappraisal of problems, and the formation of opinions about how problems might be solved are central to the understanding of history and of historical geography. Perhaps the greatest advance in historical geography in recent years has been achieved by viewing the past through the eyes of contemporary observers and by rediscovering the evaluations they made of the objects they observed.

Before the ideas and attitudes of past generations can be comprehended the material environments in which people moved and worked must be recreated.

Macaulay reminded his readers never to forget that the England of 1685 "was a very different country from that in which we live" (Macaulay, 1848, p. 279). Only by making a deliberate effort of the imagination is it possible for a northerner to visualize the Mediterranean world of classical antiquity and images of more remote ages are increasingly difficult to recall. For prehistoric times there is too little evidence to enable the true history of neolithic Liguria to be recaptured, as Croce intended, by becoming in the mind a neolithic Ligurian (Croce, 1917). Unless some expression of their abstract thoughts may be deciphered there is no means of knowing whether they would have regarded themselves as neolithic or Ligurian, let alone how they viewed their surroundings.

With a repository of ideas that have been recorded and preserved, geosophy, the study of geographical knowledge, may begin to probe what Whittlesey called "Man's sense of terrestrial space" (Whittlesey, 1945, p. 2). The scope of geosophy "covers the geographical ideas, both true and false, of all manner of people—not only geographers, but farmers and fishermen, business executives and poets, novelists and painters, Bedouins and Hottentots—and for this reason it necessarily has to do in large degree with subjective conceptions" (Wright, 1947, p. 12). From the time of the Greeks a multitude of observations and commentaries take divergent paths, some of which may be followed in *Traces on the Rhodian Shore* (Glacken, 1967), but for the most part the vast forest of "facts, lore, musings and speculations which we call the thought of an age or of a cultural tradition" (Glacken, 1967, p. viii) is largely uncharted. Ideas of cosmology and geography formulated during one of the most active, creative, and puzzling periods in the development of western thought are critically examined in *Geographical Lore of the Time of the Crusades* (Wright, 1925). In every age the views that men and women have formed of the lands they have visited or occupied have been influenced as much by their beliefs as by facts. Few studies have so completely entered into the minds of past observers as *Mirror for Americans: Likeness of the Eastern Seaboard 1810* (Brown, 1943). It faithfully reproduced an image of the United States as it might have appeared to an imaginary Philadelphian, Thomas P. Keystone, the arrangement of the book and the discussion of the evidence being put in Keystone's own words. *Historical Geography of the United States* (Brown, 1948) is also based on original eyewitness accounts but the beliefs and aims of contemporary writers are weighed and criticized in the light of modern knowledge.

When studying the infinitely divergent currents of cultural history, whether aesthetic, religious, political, social, or economic, it is impossible to avoid the conclusion that the same observed facts are capable of being arranged in many different patterns and "have different meanings to people of different cultures, or at different stages in the history of a particular culture" (Kirk, 1963, p. 366). The questions historians and geographers address to the past, while necessarily framed in terms of their own experiences of the world in which they live, may be answered in terms that fail to make sense. They need to be rationalized within alien cultural codes. "We deal not with Culture, but with cultures," wrote Sauer (1941b, p. 24), "except in so far as we delude ourselves into thinking the world made over in our own image".

A particularly rewarding field of geosophical study is afforded in those areas where people from differing cultural backgrounds have expressed their attitudes about scenery, resources, and ways of life. Europeans reacted to tropical lands in a variety of ways: they viewed the West African coast as a white man's grave, while in the Pacific they found island paradises (Smith, 1960). The nineteenth-century debate about whether the Great Plains of North America were either a pastoral garden of the west or a Great American Desert has been pursued relentlessly in the present century (Lewis, 1962). The most penetrating study of a semi-arid environment in Australia, *Back of Bourke* (Heathcote, 1965), demonstrates that traditional concepts of restrictive resource use have repeatedly frustrated attempts to exploit evanescent gifts. Opportunism has paid the highest dividends: "just as few resources are equally available in space, so few are equally available in time and like rosebuds that smile today tomorrow may be dying" (Heathcote, 1965, p. 199). Historical perception of cultures other than our own may cause us to reflect on our attitudes towards soil erosion, water conservation, and more intractable human problems of landownership, industrial organization, and population migration.

The private realms of perception and illusion, deeply explored by artists and by psychologists, are now receiving attention from geographers (Lowenthal, 1961). Geographies perceived by children, by primitive peoples, by utopian thinkers, by novelists, and by painters are being examined (Boulding, 1956; Hall, 1959; Darby, 1948; Paterson, 1965; Lowenthal and Prince, 1965; Clark, 1949; Pevsner, 1956).

MODELS IN HISTORICAL GEOGRAPHY

The real world is amorphous chaos, history a swirling blur, until an intelligent observer imposes order upon it by identifying, classifying, arranging, and relating comprehensible sections of it. To understand the world around him the theoretician creates with words, numbers, signs, and symbols a non-existent, abstract world, and seeks to identify its counterparts in actuality.

At the most general level, the operation of historical processes is represented schematically or metaphorically by theories or hypotheses explaining evolution, diffusion, cultural change, and economic growth. Less generalized theories account for shifts in agricultural and industrial locations, for changing patterns in population movements and for the development of urban structures. General theories of this order offer broadly defined, simplistic explanations of large aggregates of phenomena, but lack precision.

The analysis of processes may be performed in strictly quantitative terms by assuming a featureless initial surface and by supplying fabricated data as operative determinants. A model may be fashioned more realistically by imposing appropriate constraints and by using empirically derived data. Simulation models of the formation of transport networks and of the growth of towns have been constructed from realistic data. Development simulated by such a model may be compared with historical records (Morrill, 1965). A variant of the method employed by econometric historians constructs an initial surface with data derived from an observed situation but supplies invented data to trace a course of development that might have been followed had processes other than those recorded been operative (Desai, 1968). Geographers have built models of this type to analyse London's importance in changing English society and its economy in the seventeenth century and to analyse the spatial dynamics of urban–industrial growth in the United States from 1800 to 1914 (Wrigley, 1967; Pred, 1967).

Deterministic models are those in which the development of systems through time and space may be completely predicted from given sets of postulates. They simulate reality most closely when the processes at work are most fully understood. By assigning correct weights to the attractive power of three parent settlements, a church, and a road, a gravity model faithfully reproduced the observed chronology of colonization and explained why colonizers entered lands of low productivity before taking up better soils (Bylund, 1960). Similar models have correctly described the advance and retreat of settlement frontiers and the emergence of hierarchies of central places (Enequist, 1960; Morrill, 1963).

Where a large number of factors act independently to bring about a result, the process may closely resemble that produced by chance. Stochastic models operate, at least partially, on chance or random factors and they produce realistic images of many situations in cultural history. Hägerstrand's study of the diffusion of agricultural innovations in the management of pasture and in the prevention of bovine tuberculosis assumed that diffusion would take a partly random course (Hägerstrand, 1953). Hägerstrand's work has been followed by a number of highly revealing analyses of the development of transport links, of towns, and of settlements in general (Kansky, 1963; Curry, 1964). In all these models the operation of random factors has played a large, and in some the largest, part.

For historical geographers probabilism raises a disturbing question. If "the development of cultural forms over space is not a haphazard process" (Harvey, 1967, p. 549), how are the combined effects of an aggregation of purposeful actions to be measured when, in sum, the result is indistinguishable from a pattern produced largely by chance? At what scales does the exercise of choice and free will and the taking of decisions affect the course of history? At what level of generalization may historical geographers justify Peter Heylyn's assertion that without history, without a study of human behaviour in action, geography "hath life and motion, but at randome and unstable"?

THE GREAT LIBERATION

In 1941 Sauer spoke of the years between the appearance of "Geography as Human Ecology" (Barrows, 1923) and *The Nature of Geography* (Hartshorne, 1939) as the period of "the Great Retreat", during which geography, geology, and history were torn apart and geographers tried to limit their field of study to exclude trespassers and to avoid trespassing in neighbouring fields. "Foreword to Historical Geography" (Sauer, 1941b) heralded the beginning of the Great Liberation. In the United States Sauer and Clark (1954) proclaimed the necessity for a genetic approach, Sauer explicitly counting historical geography a part of culture history (Sauer, 1941b, p. 9). Brown (1943) and Wright (1966) gained fresh and imaginative insights into the past through the minds of contemporary thinkers. In France, Bloch (1954) and Dion (1949) demonstrated the significance of the past in the present, while in England Hoskins (1955), Beresford (1957 and 1961) and Darby (1962a and 1962b) crossed the debatable borderland between history and geography. During the past 20 years historical geographers have not only joined with historians, archaeologists, anthropologists, and sociologists in exploring the past, but they have also set out to rediscover lost worlds in the minds of the past and have begun to visualize and create in their own minds abstract worlds of theory.

REFERENCES AND FURTHER READING

Approaches to historical geography are reviewed in:

Darby, H. C. (1953) On the relations of geography and history, *Transactions of the Institute of British Geographers*, **19**, 1–11, reprinted in Taylor, G. (Ed.) (1957) *Geography in the Twentieth Century*, Methuen, London, pp. 640–52.

Smith, C. T. (1965) Historical geography: current trends and prospects, in Chorley, R. J. and Haggett, P. (Eds.) *Frontiers in Geographical Teaching*, Methuen, London, pp. 118–43.

PAST GEOGRAPHIES

Géohistoire is introduced in:

Braudel, F. (1949) *La Méditerranée et le Monde Méditerranéen à l'Époque de Philippe II*, Colin, Paris.

Primitive landscapes are discussed in:

GRADMANN, R. (1898) *Das Pflanzenleben der schwäbischen Alb*, Schnürlen, Tübingen.

SAUER, C. O. (1925) The morphology of landscape, *University of California Publications in Geography*, **2**, 19–54, reprinted in LEIGHLY, J. (Ed.) (1963) *Land and Life: a selection from the writings of Carl Ortwin Sauer*, U. of California P., Berkeley, pp. 315–50.

SAUER, C. O. (1941a) The personality of Mexico, *Geographical Review*, **31**, 353–64, reprinted in LEIGHLY, J. (Ed.) (1963) *Land and Life: a selection from the writings of Carl Ortwin Sauer*, U. of California P., Berkeley, pp. 104–17.

The nature of cross-sections is examined in:

HARTSHORNE, R. (1959) *Perspective on the Nature of Geography*, Rand McNally, Chicago.

MACKINDER, H. J. (1930) The content of philosophical geography, *International Geographical Congress, Cambridge, 1928: report of the proceedings*, Cambridge U.P., Cambridge.

An example of a reconstruction of a past geography is:

DARBY, H. C. (1952) *The Domesday Geography of Eastern England*, Cambridge U.P., Cambridge. Later volumes of this work deal with other parts of England.

GEOGRAPHICAL CHANGE

The genetic approach is discussed in:

CLARK, A. H. (1954) Historical geography, in JAMES, P. E. and JONES, C. F. (Eds.) *American Geography: inventory and prospect*, Syracuse U.P., Syracuse, pp. 70–105.

The idea of sequent occupance is put forward in:

WHITTLESEY, D. (1929) Sequent occupance, *Annals of the Association of American Geographers*, **19**, 162–5.

An essay on the significance of the frontier in American history is included in:

EDWARDS, E. E. and MOOD, F. (Eds.) (1938) *The Early Writings of Frederick Jackson Turner*, Wisconsin U.P., Madison.

The tracing of changes in individual landscape features is exemplified in:

DARBY, H. C. (1951) The changing English landscape, *Geographical Journal*, **117**, 377–98.

HELMFRID, S. (Ed.) (1961) Morphogenesis of the agrarian cultural landscape, *Geografiska Annaler*, **43**, 1–328.

MARSH, G. P. (1864) *Man and Nature; or physical geography as modified by human action*, Scribner, New York; a new edition with introduction by Lowenthal, D., Harvard U.P., Cambridge, Mass., 1965.

THOMAS, W. L. (Ed.) (1956) *Man's Role in Changing the Face of the Earth*, Chicago U.P., Chicago.

Geographical change through time is considered in:

CLARK, A. H. (1962) The sheep/swine ratio as a guide to a century's change in the livestock geography of Nova Scotia, *Economic Geography*, **38**, 38–55.

WEAVER, J. C. (1954) Changing patterns of cropland use in the Middle West, *Economic Geography*, **30**, 1–47.

WEBB, J. W. (1963) The natural and migrational components of population changes in England and Wales, 1921–1931, *Economic Geography*, **39**, 130–48.

THE PAST IN THE PRESENT

The retrogressive method is examined in:

BAKER, A. R. H. (1966) Field systems in the Vale of Holmesdale, *Agricultural History Review*, **14**, 1–24.

BAKER, A. R. H. (1968) A note on the retrogressive and retrospective approaches in historical geography, *Erdkunde*, **22**, 243–4.

BLOCH, M. (1966) *French Rural History*, Routledge & Kegan Paul, London; a translation of *Les Caractères Originaux de l'Histoire Rurale Française*, Oslo, 1931.

GULLEY, J. L. M. (1960) The Wealden Landscape in the Early Seventeenth Century and its Antecedents, unpublished Ph.D. thesis, University of London.

MAITLAND, F. W. (1897) *Domesday Book and Beyond*, Cambridge U.P., Cambridge; new edition Fontana, London, 1960.

The retrospective approach to relict features in the present landscape is discussed in:

OGILVIE, A. G. (1952) The time-element in geography, *Transactions of the Institute of British Geographers*, **18**, 1–15.

SAUER, C. O. (1941b) Foreword to historical geography, *Annals of the Association of American Geographers*, **31**, 1–24, reprinted in LEIGHLY, J. (Ed.) (1963) *Land and Life: a selection from the writings of Carl Ortwin Sauer*, U. of California P., Berkeley, pp. 351–79.

THROWER, N. J. W. (1966) *Original Survey and Land Subdivision: a comparative study of the form and effect of contrasting cadastral surveys*, Rand McNally, Chicago.

WATSON, J. W. (1959) Relict geography in an urban community: Halifax, Nova Scotia, in MILLER, R. and WATSON, J. W. (Eds.) *Geographical Essays in Memory of Alan G. Ogilvie*, Nelson, London, pp. 110–43.

YATES, E. M. (1960) History in a map, *Geographical Journal*, **126**, 32–51.

PERCEPTION OF THE PAST

The rôle of the imagination in reconstructing the past is discussed in:

CROCE, B. (1917) *Teoria e Storia della Storiografia* Bari, English translation in *Theory and History of Historiography*, Harrap, London, 1921.

MACAULAY, T. B. (1848) *The History of England from the Accession of James the Second*, Longmans, London; the third chapter described the state of England in 1685.

The geographical knowledge of the past is critically examined in:

GLACKEN, C. (1967) *Traces on the Rhodian Shore*, U. of California P., Berkeley.

WHITTLESEY, D. (1945) The horizon of geography, *Annals of the Association of American Geographers*, **35**, 1–36.

WRIGHT, J. K. (1925) *The Geographical Lore of the Time of the Crusades*, American Geographical Society, New York, reprinted 1965, Dover, New York.

WRIGHT, J. K. (1947) *Terrae Incognitae*: the place of imagination in geography, *Annals of the Association of American Geographers*, **37**, 1–15.

The past viewed through the eyes of contemporary observers is considered in:

BROWN, R. H. (1943) *Mirror for Americans: likeness of the eastern seaboard 1810*, American Geographical Society, New York.

BROWN, R. H. (1948) *Historical Geography of the United States*, Harcourt Brace, New York.

HEATHCOTE, R. L. (1965) *Back of Bourke: a study of land appraisal and settlement in semi-arid Australia*, Melbourne U.P., Melbourne.

KIRK, W. (1963) Problems of geography, *Geography*, **48**, 357–71.

LEWIS, G. M. (1962) Changing emphasis in the description of the natural environment of the American Great Plains area, *Transactions of the Institute of British Geographers*, **30**, 75–90.

SMITH, B. (1960) *European Vision and the South Pacific, 1768–1850*, Oxford U.P., Oxford.

Private realms of perception are examined in:

BOULDING, K. E. (1956) *The Image*, Michigan U.P., Ann Arbor.

CLARK, K. (1949) *Landscape into Art*, Pelican, Harmondsworth.

DARBY, H. C. (1948) The regional geography of Thomas Hardy's Wessex, *Geographical Review*, **38**, 426–43.

HALL, E. T. (1959) *The Silent Language*, Doubleday, New York.

LOWENTHAL, D. (1961) Geography, experience and imagination: towards a geographical epistemology, *Annals of the Association of American Geographers*, **51,** 241–60.

LOWENTHAL, D. and PRINCE, H. C. (1965) English landscape tastes, *Geographical Review*, **55,** 186–222.

PATERSON, J. H. (1965) The novelist and his region: Scotland through the eyes of Sir Walter Scott, *Scottish Geographical Magazine*, **81,** 146–52.

PEVSNER, N. (1956) *The Englishness of English Art*, Architectural P., London.

MODELS IN HISTORICAL GEOGRAPHY

Models concerned with change through time include:

DESAI, M. (1968) Some issues in econometric history, *Economic History Review, Second Series*, **21,** 1–16.

MORRILL, R. L. (1965) Migration and the growth of urban settlement, *Lund Studies in Geography, series B*, **26.**

PRED, A. (1967) *The Spatial Dynamics of United States Urban-industrial Growth, 1800–1914: theoretical and interpretative essays*, M.I.T.P., Cambridge, Mass.

WRIGLEY, E. A. (1967) A simple model of London's importance in changing English society and economy, *Past and Present*, **37,** 44–70.

Deterministic models are exemplified in:

BYLUND, E. (1960) Theoretical considerations regarding the distribution of settlement in inner north Sweden, *Geografiska Annaler*, **42,** 225–31.

ENEQUIST, G. (1960) Advance and retreat of rural settlement in northwestern Sweden, *Geografiska Annaler*, **42,** 211–20.

MORRILL, R. L. (1963) The development and spatial distribution of towns in Sweden: an historical-predictive approach, *Annals of the Association of American Geographers*, **53,** 1–14.

Stochastic models are exemplified in:

CURRY, L. (1964) The random spatial economy: an exploration in settlement theory, *Annals of the Association of American Geographers*, **54,** 138–46.

HÄGERSTRAND, T. (1953) *Innovationsförloppet ur korologisk synpunkt*, Gleerup, Lund; summarized in The propagation of innovation waves, in WAGNER, P. L. and MIKESELL, M. W. (Eds.) (1962) *Readings in Cultural Geography*, Chicago U.P., Chicago, pp. 355–68.

HARVEY, D. (1967) Models of the evolution of spatial patterns in human geography, in CHORLEY, R. J. and HAGGETT, P. (Eds.) *Models in Geography*, Methuen, London, pp. 549–608.

KANSKY, K. J. (1963) Structure of transportation networks; relationships between network geometry and regional characteristics, *University of Chicago Geography Research Paper*, **84.**

THE GREAT LIBERATION

The period of the Great Retreat extended between:

BARROWS, H. H. (1923) Geography as human ecology, *Annals of the Association of American Geographers*, **13,** 1–14.

HARTSHORNE, R. (1939) *The Nature of Geography*, Association of American Geographers, Lancaster, Pennsylvania.

Recent advances are indicated in:

BERESFORD, M. W. (1957) *History on the Ground*, Lutterworth, London.

BERESFORD, M. W. (1961) *Time and Place*, Leeds U.P., Leeds.

BLOCH, M. (1954) *The Historian's Craft*, Manchester U.P., Manchester.

DARBY, H. C. (1962a) The problem of geographical description, *Transactions of the Institute of British Geographers*, **30,** 1–14.

DARBY, H. C. (1962b) Historical geography, in FINBERG, H. P. R. (Ed.) *Approaches to History*, Routledge & Kegan Paul, London.

DION, R. (1949) La géographie humaine rétrospective, *Cahiers Internationaux de Sociologie*, **6,** 3–27.

HOSKINS, W. G. (1955) *The Making of the English Landscape*, Hodder & Stoughton, London.

WRIGHT, J. K. (1966) *Human Nature in Geography*, Harvard U.P., Cambridge, Mass.

CHAPTER 12

THE GEOGRAPHY
OF RURAL SETTLEMENTS

A. R. H. BAKER

APPROACHES TO THE STUDY OF RURAL SETTLEMENTS

The geographical study of rural settlements began with Ritter's work in the early nineteenth century. Since then both the content and the methodology of this study have been developed principally within a West European, particularly a German and French, context. Ritter's theme of interdependence among the elements of a landscape gave a broad base to early settlement geography, which included studies of such phenomena as rural housetypes, settlement patterns, and colonization as the results of complex man–land relationships. Such an approach has been expressed not only in the three principal general works on settlement geography (Houston, 1953; Mendöl, 1963; Schwartz, 1961) but also in syntheses of human geography, like those of Brunhes (1925) and Jones (1964). Such breadth of approach has resulted in confusion according to Stone, who offered a more restricted definition of rural settlement geography as "the description and analysis of the distribution of buildings by which people attach themselves to the land for the purposes of primary production" (Stone, 1965, p. 347). The specific restriction to buildings is made because geographical literature has become so large and research methods so specialized that in Stone's view particularizing can be afforded—in fact is needed—for ease of approach to the work and clarity of theoretical and applied results. When an explanation of the distributional pattern of buildings requires an examination of such phenomena as building materials, architectural styles, land uses, and fence types, then these elements become part of the geography of rural settlement. This narrower view has been challenged (Gordon, 1966) but its focus on buildings does provide a clearly definable core for a part of geography which is at present inadequately studied and poorly understood (Stone, 1966).

Compared with rural land use, rural settlement patterns are both much less varied regionally and much more resistant to change. There are, after all, only two basic types of rural settlement, grouped and dispersed. In the former, the population and settlement of a parish or township are concentrated at one point and there is a clear separation between the farmstead and the farm holding; in the latter, a pattern of isolated farmsteads eliminates the distinction between place of habitation and place of work, each farmstead being surrounded by its holding. But there exists, in addition, a wide range of settlement types between these two extremes. Many studies of rural settlements have elucidated the contrasting patterns to be found within particular regions, and have tended to observe these patterns as static distributions. Other studies have emphasized temporal rather than—or as well as—spatial variations in rural settlement patterns. While it is true that changes in land use and economy have often left rural settlement patterns intact and maladjusted to the new circumstances, it is also the case that these patterns have themselves evolved and at times been changed in a revolutionary manner. To the regional diversity of rural settlement patterns identified by such early workers as Meitzen (1895) have been added the temporal changes in these patterns observed by such later writers as Beresford (1954).

Geographers have studied rural settlements at a number of levels. At the lowest level, they have been interested in the characteristics and qualities of discrete structures, such as farmsteads, fences, and fields (e.g. Baker, 1965; Mead, 1966). At another level, they have been concerned with the functional interrelationships of these structures that give rise to contrasting and changing patterns of rural settlements (Gleave, 1962; Roden and Baker, 1966). The rural settlement geographer's early concern with "form, function and genesis" paralleled that of the geomorphologist's with "structure, process and stage".

RURAL SETTLEMENT STRUCTURES

Most of the earliest work in settlement geography was carried out by Germans on two main subjects: house types (including their distribution, architecture, and building materials) and urban centres. Much early work in France was also concerned with rural house forms and building materials—with rural settlements as individual structures. This is a field of study which has been extended to other parts of the world and which still has its practitioners. Nevertheless, to the study of individual buildings was soon added an examination of the relationship of one building to another—of settlement patterns.

In 1895 Meitzen's four-volume classic summarized work from 1768 on European settlement forms, classified the settlements of Germany, and established the emphasis on the form of villages which has characterized many European studies of rural settlement geography since. Most of the many attempts to classify rural settlement patterns in Europe have employed morphological criteria. Thus Demangeon (1920 and 1939) used morphology as the fundamental criterion because it was a summation of the site, function, size, and origin of a settlement. He classified French grouped rural settlements into linear, massed, and star-shaped villages, dispersed settlements into linear, nebular, hamlet, and completely dispersed. He also published one of the basic studies in rural settlement geography, in which he covered major parts of the world, probably for the first time in one publication, by describing the principal distributional patterns of rural houses (Demangeon, 1927). His classification included the relative location of farm dwellings to each other, from dispersed through intercalated to agglomerated, as well as to their fields. Other studies have used such criteria as the size, degree of compactness, shape, and regularity of settlements in attempts to produce regional classifications of rural settlement types.

The subjective nature of many of these classifications limits their utility for comparative purposes, and a basic weakness is their emphasis on morphology. The fact that settlements may be similar in form but different in origin and function renders these classifications purely descriptive. Demangeon was aware of this weakness and he himself pioneered the statistical definition of rural settlement patterns. Numerous attempts were made during the 1930's to devise statistical formulae which were precise descriptions of the relative degrees of population nucleation and dispersion in a rural community (summarized in Houston, 1953, pp. 81–5). Nevertheless, none of the statistical methods which have so far been used to describe European rural settlement patterns has been widely adopted. Houston has asserted that the attempt to reveal by statistical methods the highly variable factors influencing settlement is perhaps illusory, since the plan and pattern of regional settlement types are so variable in character that statistical analysis produces generalizations which are only partially applicable to real situations (Houston, 1953, p. 84). Many of the pre-war statistical analyses encountered considerable problems in dealing simultaneously with two distinctive characteristics of settlements, their spacing and size. Newly devised mathematical techniques, many of them (like "nearest-neighbour analysis") borrowed from the plant ecologists, are now being tested in a renewed attempt to provide mathematically precise and objective descriptions of settlement patterns (Birch, 1967; Dacey, 1962 and 1968). So far, considerably more progress has been made in the analysis of the spatial relationships of rural settlements than has been achieved in the study of the relative numbers of settlements of specific population sizes (Haggett, 1965, pp. 106–7).

RURAL SETTLEMENT ORIGINS

Classification of regional patterns of settlement structures has been seen as a necessary preliminary to an examination of their origins. According to

Houston: "The delimitation of settlement patterns and the explanation of their origin is the chief task of the student of rural settlement" (Houston, 1953, p. 80). More recently, Gordon has claimed that the concern of settlement geography should be "the study of the form of the cultural landscape, involving its orderly description and attempted explanation" (Gordon, 1966, p. 27). This genetic approach to settlement studies was pioneered by Meitzen, who attempted to demonstrate the existence in Europe of three ethnically based zones of settlement: grouped settlements and common field systems associated with Germanic conquest; isolated farmsteads and farming in severalty associated with the Celts; and round and street village forms associated with the Slavs. Such a broad classification was immediately criticized and subsequent analysis has shown his thesis to be untenable. There are, for example, many regions of scattered settlement within the area of Teutonic occupation and village settlement also occurs widely in areas in which German influence cannot easily be traced. Meitzen's hypothesis was followed by attempts to interpret the distribution of settlement in other than ethnic terms and terrain, water supply, and defence were often cited as critical genetic factors. Again, subsequent studies have shown that these factors all play their part, but they operate in the general context of the social framework of rural society and in the particular context of the relationship of settlement to agriculture (Smith, 1967). It is now clear that, in various historical periods and in differing places, the growth of population has resulted in the emergence of grouped settlements as hamlets of small agnatic groups were transformed into villages of larger agrarian societies. If the antiquity of the village form is still sought, then there is increasing evidence of grouped settlements not only in Roman but even in Iron Age times.

   The search for the origins of individual rural settlement forms has, however, been forced to place too much emphasis on the stability and continuity of regional settlement patterns through time and has tended to stress mono-causal explanations. In the final analysis, such hypotheses are in any case unverifiable and in a sense unprofitable. Even historians are today turning their attentions away from the origins of phenomena and concentrating on their functional interrelationships at a given period. Similarly, geographers now increasingly endeavour to interpret rural settlement forms as a function of agrarian, social, and economic relationships in a particular environment at any given time period. Smith, among others, strongly advocated this functional approach to the historical geography of rural settlement and we can be certain that it is in this direction that the subject is most likely to advance (Smith, 1965 and 1967).

RURAL SETTLEMENT FUNCTIONS

Despite the pleadings of such American geographers as Hartshorne (1959, p. 107) and Gordon (1966), it is clear that studies of rural settlements will in future pay increasing attention to what geomorphologists refer to as "process". In one sense, this will involve a renewed emphasis on what Brunhes (1925) termed *connexité* and Sauer (1925) saw as the interrelationships of objects which exist together in the landscape. This approach will see the extension to studies of rural settlements of the methods of systems analysis (Foote and Greer-Wootten, 1968), with their stress on the measurement and examination of sets of objects (in this instance, rural settlements) interrelated through circulating movements (money, migrants, products, ideas, etc.). Closer study can be expected, for instance, of those influences which tend towards the grouping or dispersion of settlement. In this light, terrain, soils, water supply, availability of pasture, degree of collective organization, the needs of defence, the strength of manorial authority, etc., are seen as some of the physical, biological, and cultural inputs into the settlement system which have, often in isolation or only in limited interrelation, already been the basis of historical studies of the genesis of rural settlements.

A number of writers have recently stressed the purely locational relationships of rural settlements, working on the premises that the spatial distribution of human activity reflects an ordered adjustment to the factor of distance, that locational decisions are generally taken so as to minimize movements, and that all locations are endowed with a degree of accessibility but some locations are more accessible than others (Garner, 1967). Thus Haggett (1965) has pointed out that the traditional requirements of a village—land, water, building materials, fuel, etc.—will exert varying pulls on the location of the settlement, distorting a theoretically regular distribution of settlements.

In Fig. 11a, seven settlements are distributed regularly over an area of uniform resources, exemplifying a pattern of dispersed farmsteads. In Fig. 11b, a zonal resource (stippled) is introduced (such as a region of pasture or woodland). On the assumption that all settlements require access to this resource but that they will move as short a distance as possible from their ideal positions in a regular distribution, a new set of locations is established, exemplifying a pattern of "waste-edge" settlements such as have been observed around Dartmoor and the Weald.

In Fig. 11c, a linear resource, such as a river or road, requires a new distribution of settlements if each settlement is to have access to the resource: the resulting string-like settlements resemble the forest, marsh, and street villages of Europe. In Fig. 11d, a point resource is assumed (e.g. single well or defensive site) and the appropriate changes in location are made, exemplifying

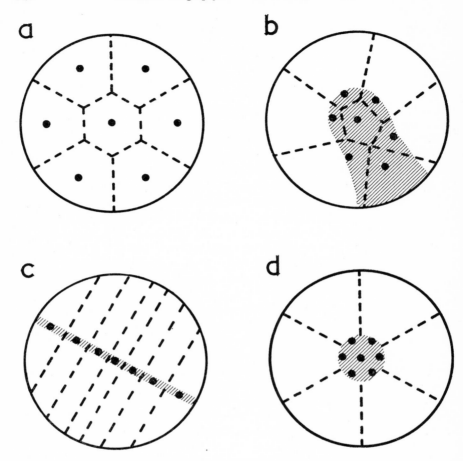

FIG. 11. Some theoretical rural settlement patterns associated with an increasingly localized resource. (After Haggett, P., 1965, p. 95.)

the round village with its radial fields and the grouped village with its three, four, five, or (as in Fig. 11d) six fields. As Haggett concluded:

> Clearly the actual development of regional settlement patterns is a multivariate product in which social conventions play as big a part as environment. Nevertheless, basic geometrical considerations, even though severely modified, still play a basic part in that syndrome [Haggett, 1965, p. 95].

Both the grouped and dispersed settlement patterns have been discussed theoretically in terms of the principle of economy of movement in farming operations, i.e. the farmer's desire to minimize the distances he needs to travel to the site of labour, or the distances over which he needs to herd stock, to move farm implements, or to gather in crops (Orwin and Orwin, 1954; Perpillou, 1966; Smith 1967). Under the farming system in which three open fields are cultivated in common, the optimum settlement form, whatever its shape, would be a nucleated village located at the central point of the three fields. If the three fields were roughly equal in area, this would, in fact, be at the centre of gravity for the cultivated areas as a whole. Where holdings were compact, it would have been logical to place the farmstead more or less centrally in order to minimize the movement of men and animals. The creation of new settlements may also be seen as an attempt to minimize movements; a moment must have arrived when, rather than return each evening a great distance to the village, peasant farmers constructed a shelter which later became a permanent site of occupation (Orwin and Orwin, 1954, p. 162; Duby, 1968). Distance factors in rural settlement patterns have been discussed by Chisholm, who suggested that:

> A very fruitful method of classifying settlement patterns could be based upon the distance between farmstead and field, which would provide material invaluable for planning and development and significant in describing the relationships between man and his physical environment probably more useful than is vouchsafed by the currently popular methods of measuring the degree of dispersion and nucleation [Chisholm, 1962, p. 126].

Rural settlement systems may be seen as adjustments to a mixture of physical, biological, and cultural inputs, so that a change in one or more of the inputs may be expected to lead to a readjustment within the system. Numerous historical and geographical studies have been concerned with tracing the evolution of settlement patterns through time and have drawn upon a wide range of evidence. In addition, a beginning has now been made towards the elaboration of theoretical models for the study of the expansion of settlement through time. Some are deterministic models (e.g. Bylund, 1960), which make certain assumptions about the nature of new colonization (for example, that the physical conditions of the land are equal in all areas and that further areas will not be settled until those close to the "mother settlements" have been occupied). Others are probabilistic models (e.g. Morrill, 1962), in which the growth of settlement patterns is simulated by random processes which are in turn restricted by the operation of certain rules based on empirical observations of colonization processes. But this is no more than a beginning and the application of models of this type seems to be greatly limited by the difficulty and sometimes impossibility of gauging the approximation of the theoretical patterns to real systems (Harvey, 1967).

CONCLUSION

The foregoing discussion is based on the view of man as an economically motivated and rational being. It has been assumed that individuals or groups arrange themselves spatially so as to optimize their given set of resources and demands. Discrepancies between the ideal and the observed patterns of rural settlement may simply reflect the fact that much human behaviour is of a sub-optimal nature (Pred, 1967), since both individuals and groups are often satisfied with less than the ideal. Frequently the information at their disposal is inadequate to make economic or rational decisions. In consequence, studies of settlements are turning increasingly to probability theory, with its emphasis on the role of choice and of chance (Curry, 1964).

Applications of systems analysis and probability theory are currently changing the geographer's attitude to settlement studies, as they are changing his attitudes to most problems in human and physical geography. To take but one example: it is now accepted that many processes can achieve a "steady state". This state can be reached regardless of any initial conditions or origins but, once established, fluctuations occur constantly in such a manner that the state is maintained. Thus, for a system in a steady state, there is no need to study its history in order to understand its form or the processes going on. Such an approach does not, according to Curry (1966), fit well the traditional trilogy of form, genesis, and function as the approaches to geographical knowledge.

Three basic attitudes may be adopted towards a study of rural settlement. The first may be termed "retrogressive", with its interest lying in the past and its consideration of the present pattern of rural settlement being concerned with the light which this throws on earlier conditions. The second may be termed the "retrospective" approach, with its focus of interest on the present, the past being considered in so far as it furthers an understanding of the present (Baker, 1968). The third may be termed the "prospective" approach, with its concern being for the future, past and present settlement forms being regarded as relict features out of adjustment with future probable needs (Juillard, 1964). Clearly these three attitudes are complementary. Until recently, human geography has been principally retrogressive and retrospective in viewpoint, with an express emphasis on the historical background of social behaviour. In that probability analysis emphasizes choice and decisions and thus looks forwards as readily as backwards, it is to be expected that its methods will be increasingly adopted in a prospective approach to the geographical study of rural settlements.

REFERENCES AND FURTHER READING

APPROACHES TO THE STUDY OF RURAL SETTLEMENTS

BAKER, A. R. H. (1965) Field patterns in seventeenth century Kent, *Geography*, **50,** 18–30.
BERESFORD, M. W. (1954) *The Lost Villages of England*, Lutterworth, London.
BRUNHES, J. (1925) *La Géographie Humaine*, 3rd edn., Libraire Félix Alcan, Paris.

GLEAVE, M. B. (1962) Dispersed and nucleated settlement in the Yorkshire Wolds 1770–1850, *Transactions of the Institute of British Geographers*, **30**, 105–18.

GORDON, T. G. (1966) On the nature of settlement geography, *Professional Geographer*, **18**, 26–8.

HOUSTON, J. M. (1953) *A Social Geography of Europe*, Duckworth, London.

JONES, E. (1964) *Human Geography*, Chatto & Windus, London.

MEAD, W. R. (1966) The study of field boundaries, *Geographische Zeitschrift*, **54**, 101–17.

MEITZEN, A. (1895) *Siedlung und Agrarwesen der Westgermanen und Ostgermanen, der Kelten, Romer, Finnen und Slawen* Berlin, 3 vols. and atlas.

MENDÖL, T. (1963) *Általános Településfoldrajz*, Budapest.

RODEN, D. and BAKER, A. R. H. (1966) Field systems of the Chiltern Hills and of parts of Kent from the late thirteenth to the early seventeenth century, *Transactions of the Institute of British Geographers*, **38**, 73–88.

SCHWARTZ, G. (1961) *Allgemeine Siedlungsgeographie*, Walter de Gruyter, Berlin.

STONE, K. H. (1965) The development of a focus for the geography of settlement, *Economic Geography*, **41**, 346–55.

STONE, K. H. (1966) Further development of a focus for the geography of settlement, *Professional Geographer*, **18**, 208–9.

RURAL SETTLEMENT STRUCTURES

BIRCH, B. P. (1967) The measurement of dispersed patterns of settlement, *Tijdschrift voor Economische en Sociale Geografie*, **58**, 68–75.

DACEY, M. F. (1962) Analysis of central place and point patterns by a nearest neighbour method, *Lund Studies in Geography, series B*, **24**, 55–75.

DACEY, M. F. (1968) An empirical study of the areal distribution of houses in Puerto Rico, *Transactions of the Institute of British Geographers*, **45**, 51–69.

DEMANGEON, A. (1920) L'habitation rurale en France, *Annales de Géographie*, **29**, 352–75.

DEMANGEON, A. (1927) La géographie de l'habitat rural, *Annales de Géographie*, **36**, 1–23 and 97–144. Part of this paper has been translated and published as: The origins and causes of settlement types, in WAGNER, P. L. and MIKESELL, M. W. (Eds.) (1962) *Readings in Cultural Geography*, U. of Chicago P., Chicago, pp. 506–16.

DEMANGEON, A. (1939) Types de villages en France, *Annales de Géographie*, **48**, 1–21.

DICKINSON, R. E. (1949) Rural settlements in the German lands, *Annals of the Association of American Geographers* **39**, 239–63.

HAGGETT, P. (1965) *Locational Analysis in Human Geography*, Arnold, London.

PFEIFER, G. (1956) The quality of peasant living in Central Europe, in THOMAS, W. L. (Ed.) *Man's Role in Changing the Face of the Earth*, U. of Chicago P., Chicago, pp. 240–77.

SHEPPARD, J. A. (1966) Vernacular buildings in England and Wales: a survey of recent work by architects, archaeologists, and social historians, *Transactions of the Institute of British Geographers*, **40**, 21–37.

THORPE, H. (1961) The green village as a distinctive form of settlement on the North European Plain, *Bulletin de la Société Belge d'Etudes Géographiques*, **30**, 93–133.

THORPE, H. (1964) Rural settlement, in WATSON, J. W. and SISSONS, J. B. (Eds.) *The British Isles, A systematic geography*, Nelson, London.

TREWARTHA, G. T. (1962) Types of rural settlement in colonial America, in WAGNER, P. L. and MIKESELL, M. W. (Eds.) *Readings in Cultural Geography*, U. of Chicago P., Chicago, pp. 517–38.

WAGSTAFF, J. M. (1965) House types as an index in settlement study: a case study from Greece, *Transactions of the Institute of British Geographers*, **37**, 69–75.

RURAL SETTLEMENT ORIGINS

JOHNSON, J. H. (1961) The development of the rural settlement pattern of Ireland, *Geografiska Annaler*, **43**, 165–73.

SMITH, C. T. (1965) Historical geography: current trends and prospects, in CHORLEY, R. J. and HAGGETT, P. (Eds.) *Frontiers in Geographical Teaching*, Methuen, London, pp. 118–43.
SMITH, C. T. (1967) *An Historical Geography of Western Europe Before 1800*, Longmans, London.
UHLIG, H. (1961) Old hamlets with infield and outfield systems in western and central Europe, *Geografiska Annaler*, **43**, 285–312.

RURAL SETTLEMENT FUNCTIONS

BRACEY, H. E. (1962) English central villages: identification, distribution and functions, *Lund Studies in Geography, series B*, **24**, 169–90.
BYLUND, E. (1960) Theoretical considerations regarding the distribution of settlement in inner north Sweden, *Geografiska Annaler*, **42**, 225–31.
CHISHOLM, M. (1962) *Rural Settlement and Land Use*, Hutchinson, London.
CLARKE, D. (1968) *Analytical Archaeology*, Methuen, London. Includes a useful discussion of systems analysis and models.
DUBY, G. (1968) *Rural Economy and Country Life in the Medieval West*, Arnold, London.
FOOTE. D. C. and GREER-WOOTTEN, B. (1968) An approach to systems analysis in cultural geography, *Professional Geographer*, **20**, 86–91.
GARNER, B. (1967) Models of urban geography and settlement location, in CHORLEY, R. J. and HAGGETT, P. (Eds.) *Models in Geography*, Methuen, London, pp. 303–60.
HARTSHORNE, R. (1959) *Perspective on the Nature of Geography*, Rand McNally, Chicago.
HARVEY, D. (1967) Models of the evolution of spatial patterns in human geography, in CHORLEY, R. J. and HAGGETT, P. (Eds.) *Models in Geography*, Methuen, London, pp. 549–608.
MORRILL, R. L. (1962) Simulation of central place patterns over time, *Lund Studies in Geography, series B*, **24**, 109–20.
ORWIN, C. S. and ORWIN, C. S. (1954) *The Open Fields*, Clarendon P., Oxford.
PERPILLOU, A. (1966) *Human Geography*, Longmans, London.
SAUER, C. (1925) The morphology of landscape, *University of California Publications in Geography*, **2**, 19–54, reprinted in LEIGHLY, J. (Ed.) (1963) *Land and Life: a selection from the writings of Carl Ortwin Sauer*, U. of California P., Berkeley, pp. 104–17.

CONCLUSION

BAKER, A. R. H. (1968) A note on the retrogressive and retrospective approaches in historical geography, *Erdkunde*, **22**, 243–4.
CURRY, L. (1964) The random spatial economy: an exploration in settlement theory, *Annals of the Association of American Geographers*, **54**, 138–46.
CURRY, L. (1966) Chance and landscape, in HOUSE, J. W. (Ed.) *Northern Geographical Essays in Honour of G. H. J. Daysh*, U. of Newcastle-upon-Tyne P., Newcastle, pp. 40–55.
JUILLARD, E. (1964) Géographie rurale française. Travaux récents (1957–63) et tendances nouvelles, *Etudes Rurales*, **13–14**, 46–70.
PRED, A. (1967) Behaviour and location, *Lund Studies in Geography, series B*, **27**, 1–128.

CHAPTER 13

THE ECOLOGY
OF AGRICULTURAL SYSTEMS

D. R. HARRIS

AGRICULTURAL GEOGRAPHY IN RETROSPECT

The traditional approach to the geographical study of agriculture in schools and universities, particularly in Britain, focused on the world production and exchange of agricultural commodities. It was epitomized by that fat and famous tome, *Chisholm's Handbook of Commercial Geography*, which was first published in 1889 and reached its eighteenth edition in 1966. The first moves away from stereotyped textbook description of production and trade—which still finds its schoolroom echo in sketch maps of the "wheat-on-the-prairies" and "cattle-on-the-pampas" type—were made in the 1920's and 1930's. During those decades interest shifted from the study of individual commodities to the areal study of agriculture on both large and small scales.

This development took place first and went farthest in the United States, where geographers sought increasingly to understand the physical and socio-economic conditions governing agricultural production in specific areas. Particular emphasis was laid on the detailed field study of small areas and the recording of agricultural land use, soil, and slope on large-scale maps (Jones and Sauer, 1915; Jones and Finch, 1925; Finch, 1933). This trend culminated in several nationally organized programmes of land classification and mapping, such as those of the Tennessee Valley Authority, and the Land Utilisation Survey of Great Britain. A somewhat later development was the intensive investigation of individual units of agricultural production, such as a mixed farm, a ranch, or a plantation, which were regarded as typical of a particular area or form of agriculture (Platt, 1930 and 1942; Blaut, 1953).

Attempts to categorize world agricultural regions go back into the nineteenth century, particularly in the work of German geographers (Ratzel, 1891; Hahn, 1892), and in the 1920's and 1930's American geographers showed a keen interest in agricultural distributions at both continental and world scales. O. E. Baker was the leading exponent of this approach with his *Atlas of American Agriculture* (1924) and his articles on the agricultural regions of North America published in *Economic Geography* between 1926 and 1933. The latter were paralleled by a series of papers by other geographers on the agricultural regions of the remaining continents published in the same journal between 1925 and 1943. In the 1930's, too, Derwent Whittlesey produced his well-known delineation of world agricultural regions (Whittlesey, 1936), modified versions of which are still widely used in atlases and textbooks (e.g. *Goode's World Atlas*; Jones and Darkenwald, 1965; Symons, 1969).

Through the inter-war period the selection of criteria by which agricultural regions were defined became progressively more sophisticated. The simple, often subjective data used in early schemes were replaced by more complex, quantitative measures. For example, in 1930 Jones proposed a new and more precise definition of commercial grain farming, dairy farming, and commercial livestock farming based on four ratios: the percentage of land in crops, livestock units per hundred acres in crops, the ratio of gallons of milk to acres in crops, and the percentage of total livestock represented by work animals (Jones, 1930). Attempts to refine the definition of agricultural regions continued after the Second World War, with such formulations as Weaver's statistical analyses of crop combinations in the Middle West (Weaver, 1954a, b), but by this time interest in the refractory problem of broad regionalization was fading and agricultural geography began to move at unequal pace in two new directions: on the one hand towards economics and on the other towards ecology.

The move towards a genuinely economic geography of agriculture started first and has progressed farthest. It represents an extension of the mainstream of agricultural geography and owes its success largely to the infusion of economic theory into traditional commercial geography. Its chief practitioners have come to the subject with qualifications in both economics and geography. In Britain R. O. Buchanan at the London School of Economics and W. R. Mead at University College London helped initiate this shift of emphasis and now study of the economic geography of contemporary agriculture is well established in both British and American geography departments (see Chapter 14, below).

ECOLOGY AND AGRICULTURAL GEOGRAPHY

Agricultural geography has moved more hesitantly and less far in the second new direction: towards ecology. This approach conceives of agriculture not simply in relation to, but as an integral part of, the environment in which it is practised. Crops and livestock, as well as more strictly cultural attributes such as tools and techniques of tillage, planting, and harvesting, are all regarded as components of given ecosystems. Thus agricultural *systems* are recognized simply as distinctive types of man-modified ecosystems (ecotypes), whether they are forms of traditional, "palaeotechnic" cultivation in which human and animal labour predominates, or modern, "neotechnic" farming which depends increasingly on scientific skills and the energy derived from combustible fuels.

The methodological and practical merit of this ecosystematic approach is that—as with the application of general systems theory to other aspects of geography (Stoddart, 1965)—it focuses attention on the basic properties common to all systems: structure, function, equilibrium, and change. It thus leads us to ask five fundamental questions about each agricultural system:

(1) How is it organized or what is its structure?
(2) How does it function?
(3) What degree of stability does it have?
(4) How did it evolve through time?
(5) How will it develop in the future?

Attempts to answer the first four questions are necessary before informed projections can be made about the future development of a system under a variety of postulated conditions.

Recently considerable progress has been made towards the structural and functional analysis of agricultural systems by the use of models which emphasize the economic and social attributes of agriculture (Henshall, 1967), but, although geographers have examined the relevance of ecological concepts to tropical farming (Blaut, 1961), agricultural land use (Simmons, 1966), and the origins of agriculture (Harris, 1969), few ecological analyses of specific agricultural systems have as yet been carried out.

The earliest systematic attempts to interpret the subsistence patterns of human groups in ecological terms were made by American anthropologists, first and foremost by Steward (Steward, 1955; Mikesell, 1967). He analysed comparatively the material culture and social organization of simple, mainly non-agricultural groups and tried to identify those attributes of the physical environment that are functionally related to subsistence activities. Although theoretically Steward placed the study of agricultural systems, as well as of other patterns of subsistence, firmly in the centre of his anthropological stage, he and most of his followers neglected the study of agricultural peoples in favour of non-agricultural hunting and gathering groups. Nevertheless, one or two anthropologists did attempt to modify Steward's methods of "cultural ecology" and extend them to the analysis of more complex agricultural communities. Most notable is the work of Geertz on shifting (*swidden*) cultivation and wet-padi farming in Indonesia (Geertz, 1963). He took the ecological approach a stage further than Steward by relating to the traditional agricultural systems of Southeast Asia the principles of ecosystem analysis enunciated by Odum (1959 and 1963). Wolf, too, has examined the agricultural systems of peasant communities in ecological terms (Wolf, 1966).

What is now needed is that the methods of analysis pioneered by Steward, Geertz, and others should be refined and applied to a much wider range of agricultural systems. These should include both neotechnic systems of the modern world, such as dairy farming, commercial grain farming, commercial livestock ranching, and plantation agriculture, and palaeotechnic systems of the traditional "non-Western" world, such as fixed-plot horticulture, forms of mixed grain–livestock farming, and nomadic pastoralism. A major advantage of this approach is that it provides a framework for the analysis of all agricultural systems and subsystems from the scale of a world agricultural region to that of a single farm or field, the ecological structure, functioning, and history of which could well be investigated as a school class project. By way of example, at an intermediate scale, the ecological characteristics of two of the palaeotechnic agricultural systems of tropical America will be briefly compared, but first it is necessary to examine more closely the relationships that exist between agricultural systems and *natural* ecosystems. (For a fuller discussion of the ecosystem concept see Chapter 4, above.)

Natural Ecosystems and Agricultural Systems

In the study of major natural ecosystems a fundamental distinction can be made between generalized and specialized types. The generalized ecosystems are characterized by a great variety of plant and animal species each of which is represented by a relatively small number of individual organisms. Thus the diversity index of the ecosystem—or the ratio between numbers of species and of individuals—is high. Conversely, specialized ecosystems have a low diversity index and are characterized by a small variety of species, each of which is represented by a relatively large number of individuals.

In generalized ecosystems net primary productivity, or the increment of plant material per unit of time, tends to be high, and many ecological niches are available to species at all trophic levels in the food web, from the primary producers (green plants) to the primary, secondary, and tertiary consumers (herbivores, carnivores, and top carnivores), to the decomposers (macroorganisms such as worms and wood-lice and microscopic protozoa, fungi, and bacteria). The structural and functional complexity of generalized ecosystems results in their having greater stability (homeostasis) than specialized ecosystems. Thus the reduction or removal of a component species, whether by natural or human agency, tends to have less effect because alternative pathways for energy flow are available within the system. When alternative food sources exist for many species at each trophic level, population levels fluctuate less widely and changes in one component are less likely to trigger off a chain reaction affecting the whole ecosystem. The tropical rain forest is the most highly generalized, productive, and stable of major terrestrial ecosystems. It has the highest diversity index and, although there are very few precise measurements available, net primary productivity of above-ground plant parts reaches 10–20 g/m^2 per day or 3600–7200 g/m^2 per year (Billings, 1964; Westlake, 1963).

By contrast specialized natural ecosystems are much less productive and tend also to be less stable. They include the tundra, the average annual above-ground primary productivity of which is less than 1 g/m^2 per day; the mid-latitude grasslands, whose average annual primary productivity ranges from about 0.5–2 g/m^2 per day; and the boreal forest, with average annual primary productivity of up to about 2.5 g/m^2 per day. There is not always an inverse relationship between the degree of specialization of an ecosystem and its productivity—for example, desert ecosystems are characteristically more generalized than tundra, grassland, or boreal forest ecosystems and yet their average annual primary productivity is usually less than 0.5 g/m^2 per day— but there is nevertheless a general tendency for primary productivity to be higher in the more generalized ecosystems. Certainly a gradient is apparent in major forest ecosystems from the most highly generalized and productive type, the equatorial evergreen rain forest; through the tropical seasonal semi-evergreen and deciduous forests, where species diversity is less and growth limited by a dry season the length and intensity of which increases with latitude; to the more specialized and less productive mid-latitude temperate deciduous and evergreen forests where growth is checked by winter cold.

If we extend this analysis of natural ecosystems to the interpretation of agricultural systems it is at once apparent that most neotechnic agricultural systems are highly specialized; they exist to produce maximum numbers of optimum-sized individuals of one or two preferred plant or animal species. Some palaeotechnic agricultural systems are similarly, if rather less highly, specialized. Wet-padi rice cultivation and nomadic pastoralism are both dependent on a very limited range of domestic crops and livestock and have evolved special techniques for raising them and for maintaining the productivity of the system (periodic flooding of the rice on the one hand and seasonal migration of the herds on the other). Many traditional agricultural systems, however, are more generalized. They are polycultural rather than monocultural, raising a diverse assemblage of crops in functional interdependence and sometimes integrating livestock into the system as both consumers and fertilizing agents. Swidden cultivation and fixed-plot horticulture are examples of such generalized systems still widely practised in the tropics, while in mid-latitudes mixed farming—involving the production on the same land of crop combinations of grains, roots, and livestock—represents a somewhat less generalized system which has become more specialized by reaching a higher level of technical complexity.

The Agricultural Modification of Natural Ecosystems

From this comparison between natural ecosystems and agricultural systems, four ways in which the advent of cultivation changes natural ecosystems can be deduced. The mode of change most apparent today is when generalized natural ecosystems have been transformed into specialized artificial ecosystems. This involves a drastic reduction in the diversity index following replacement of most of the wild species by a relatively small complement of cultivated plants and domestic animals. The transformation may also lead to the selective increase of certain wild species that thrive in the disturbed habitats associated with cultivation and settlement (both plant and animal "weeds"), but, by building-up their populations at the expense of more vulnerable members of the natural community, the specialized nature of the ecosystem is enhanced and the diversity index remains low.

The transformation of generalized natural ecosystems into specialized agricultural systems usually leads to a loss of net primary productivity, but this is not always so. Under modern methods of intensive farming the trend may be reversed. The productivity of sugar cane under intensive cultivation in Hawaii (6700 g/m² per year) falls within the upper range of estimates already quoted for the tropical rain forest; and that of a fertilized maize field in Minnesota was found to be approximately equal to the annual net primary productivity of a nearby deciduous oak wood which had been protected from exploitation (Ovington *et al.*, 1963).

The second mode of change—the transformation of specialized natural ecosystems into more generalized agricultural systems—has happened only rarely. The introduction into mid-latitude grasslands of a crop–livestock–weed complex associated with a system of mixed farming, as occurred, for example, during the nineteenth century on the American prairie and the Argentinian pampas, probably resulted in an increase in diversity index; as has the establishment of polycultural irrigation agriculture in certain desert ecosystems in the twentieth century. But where agriculture has intruded into specialized natural ecosystems the third mode of change has usually occurred. This involves the replacement of specialized natural ecosystems by still more specialized agricultural ones based on monoculture, such as cereal dry-farming or cotton irrigation, with a resultant lowering of the diversity index.

Lastly the agricultural use of a natural ecosystem may be accomplished by manipulation rather than transformation; not by drastically changing its diversity index, but by altering selected components without fundamentally modifying its overall structure. Instead of an artificial ecosystem being created to replace the natural one, cultivation may proceed by substituting certain preferred domesticated species for wild species that occupy equivalent ecological niches. Thus an assemblage of cultivated trees and shrubs, climbers, herbs, and root crops may take over spatial and functional roles essentially similar to those fulfilled by wild species of equivalent life-form in the natural ecosystem. Swidden cultivation and fixed-plot horticulture manipulate the generalized ecosystems of tropical forests in this way and in so doing come closer to simulating the structure, functional dynamics, and equilibrium of the natural ecosystem than any other agricultural systems man has devised.

The substitution for equivalent wild species of domestic animals rather than crops occurs less frequently in manipulated ecosystems. Certain domesticates—notably dogs and pigs—may fulfil a rôle as scavengers, but this function is normally confined to the immediate vicinity of settlements and does not lead to replacement of the more widely ranging wild scavengers. The free-range management of pigs in the forests of medieval Europe, and of cattle in pre-Roman times, may be thought of as an example of the partial substitution of domestic for wild scavengers and browsers.

TWO CASE STUDIES: SEED-CULTURE AND VEGECULTURE IN THE AMERICAN TROPICS

In conclusion, the ecological approach to agricultural geography can be exemplified by comparing briefly two of the palaeotechnic agricultural systems of tropical America. They are both examples, at an intermediate scale, of systems that manipulate rather than transform natural ecosystems.

When the Europeans discovered the New World, agriculture was practised in most of Central and South America and also in southwestern and eastern North America. In terms of crop complexes the fundamental contrast was between agricultural systems primarily dependent on seed-reproduced and vegetatively reproduced crops. "Seed-culture" was based upon a highly productive combination of maize (*Zea mays*), beans (*Phaseolus* spp.), squashes (*Cucurbita* spp.) and other seed crops such as chili peppers (*Capsicum* spp.). Its focus was in Mexico with extensions northward across the Rio Grande and southward into northern South America. It provided a uniquely well-balanced vegetable diet which required little or no supplementation with animal protein.

"Vegeculture" on the other hand was based on starch-rich root crops, such as manioc or cassava (*Manihot esculenta*), sweet potato (*Ipomoea batatas*), and arrowroot (*Maranta arundinacea*), which are propagated not from seed but by planting stem cuttings or tubers. It was most highly developed in the humid tropical lowlands of South America east of the Andes, with extensions up the east coast of Central America, and, as a temperate highland variant based on the potato (*Solanum tuberosum*), into the Andes. In the Andes, in Central America, and in the West Indies the two crop complexes met and mingled.

Both systems are still widely practised in the American tropics by the techniques of swidden cultivation. But there is a tendency for the clearings dominated by root crops, which are commonly known in Spanish-speaking South America as *conucos*, to be cultivated for longer periods than those devoted to seed crops, which are known in Middle America as *milpas*. This tendency is sometimes accentuated by the practice in conucos of planting cuttings and tubers in specially prepared earth mounds or *montones*. Both systems of cultivation are also polycultural. A variety of upright, climbing, and sprawling crops are grown in close association; in conucos they may be planted together in the same mound.

But there are significant differences in the structure and equilibrium of the two systems. The diversity of plants present tends to be greater, their stratification more complex and the canopy of vegetation more completely closed in conuco cultivation: in other words it represents a more highly generalized ecosystem than milpa cultivation. Furthermore, because conuco productivity is focused upon starchy root crops, it makes smaller demands on plant nutrients than the relatively protein-rich crops of milpa, notably maize and beans. Less fertility is therefore removed from the soil at harvest and the system has greater inherent stability than milpa. This is further enhanced by the fact that the opportunity for soil erosion is minimal because the ground is seldom bare of plant cover and, when present, the mounds effectively check sheetwash. Provided the soil is deep enough, conucos can be cultivated successfully even on steep slopes without inducing erosion. The milpa ecosystem, on the other hand, much more easily gets out of equilibrium. The pre-eminence of maize and other nutrient-demanding crops, the less complex stratification, and the more open canopy which increases opportunities for weed invasion, all combine to make milpa cultivation less conservative of soil resources and more prone to shift from one temporary clearing to another.

The instability and expansive nature of maize-dominated milpa swidden has been well demonstrated for the Maya lowland of Yucatan (Cowgill, 1962); and historically we should expect milpa to exhibit a much greater tendency than conuco to expand into new areas. This may indeed be one of the principal reasons for the predominance of seed-culture over vegeculture in the agricultural systems of aboriginal America at the time of European discovery, related as it would have been to the gradual ascendency of the seed planters' balanced vegetable diet, based on maize and its associated crops, over the root planters' starch-rich vegetable diet, dependent upon a local supply of animal protein. Thus seed-culture was not only less stable but it also had a greater inherent tendency to expand, and, by providing a balanced vegetable diet, it allowed expansion to take place into areas where little animal protein was available. Conversely, vegeculture created a more stable ecosystem and remained tied to river-bank, seashore, savanna-edge, and other habitats with assured supplies of animal protein. Similarly, in Southeast Asia, an historical pattern of seed-culture expanding into areas of vegeculture is apparent. There an intrusive rice culture has progressively replaced an indigenous vegecultural system based on yam and taro cultivation (Spencer, 1966, pp. 110–22).

These examples illustrate the value of an ecological approach to the geography of agricultural systems. It provides a mode of analysis, applicable at any scale, that deepens our understanding of agriculture both in the present and in the past. Together with more orthodox economic analysis it promises insights into the feasibility of future agricultural development. And, last but not least, it depends for its success—in teaching and research—on a fusion of strands from both the physical and the human side of our traditionally divided subject.

REFERENCES AND FURTHER READING

AGRICULTURAL GEOGRAPHY IN RETROSPECT

The following are characteristic of the areal studies that dominated agricultural geography in the first part of the twentieth century:

BIRCH, J. W. (1954) Observations on the delimitation of farming type regions, with special reference to the Isle of Man, *Transactions of the Institute of British Geographers*, **20**, 141–58.

BLAUT, J. M. (1953) The economic geography of a one-acre farm in Singapore: a study in applied microgeography, *Malayan Journal of Tropical Geography*, **1**, 37–48.

FINCH, V. C. (1933) Montfort—a study in landscape types in southwestern Wisconsin, *Bulletin of the Geographic Society of Chicago*, **9**.

HAHN, E. (1892) Die Wirtschaftsformen der Erde, *Petermann's Mitteilungen*, **38**, 8–12.

HARTSHORNE, R. and DICKEN, S. N. (1935) A classification of the agricultural regions of Europe and North America on a uniform statistical basis, *Annals of the Association of American Geographers*, **25**, 99–120.

JONES, W. D. (1930) Ratio and isopleth maps in regional investigation of agricultural land occupance, *Annals of the Association of American Geographers*, **20,** 177–95.

JONES, W. D. and FINCH, V. C. (1925) Detailed field mapping in the study of the economic geography of an agricultural area, *Annals of the Association of American Geographers*, **15,** 148–57.

JONES, W. D. and SAUER, C. O. (1915) Outline for field work in geography, *Bulletin of the American Geographical Society*, **47**, 520–5.

PLATT, R. S. (1930) Pattern of occupancy in the Mexican Laguna District, *Transactions of the Illinois State Academy of Science*, **22**, 533–41.

PLATT, R. S. (1942) *Latin America: countrysides and united regions*, McGraw-Hill, New York.

RATZEL, F. (1891) *Anthropogeographie: Vol. 2, Die geographische Verbreitung des Menschen*, Englehorn, Stuttgart.

WEAVER, J. C. (1954a) Crop-combination regions in the Middle West, *Geographical Review*, **44,** 175–200.

WEAVER, J. C. (1954b) Crop-combination regions for 1919 and 1929 in the Middle West, *Geographical Review*, **44,** 560–72.

WHITTLESEY, D. (1936) Major agricultural regions of the earth, *Annals of the Association of American Geographers*, **26**, 199–240.

Recent methodological discussion in agricultural geography is exemplified by:

BUCHANAN, R. O. (1959) Some reflections on agricultural geography, *Geography*, **44,** 1–13.

CHISHOLM, M. (1964) Problems in the classification and use of farming-type regions, *Transactions of Institute of British Geographers*, **35,** 91–103.

HENSHALL, J. D. (1967) Models of agricultural activity, in CHORLEY, R. J. and HAGGETT, P. (Eds.) *Models in Geography*, Methuen, London, pp. 425–58.

REEDS, L. G. (1964) Agricultural geography: progress and prospects, *Canadian Geographer*, **8,** 51–63.

The only recent textbook specifically on agricultural geography is:

SYMONS, L. (1966) *Agricultural Geography*, Bell, London, but the geography of world agriculture is reviewed in most economic geography texts, e.g.:

JONES, C. F. and DARKENWALD, G. G. (1965) *Economic Geography*, 3rd edn., Macmillan, New York.

ECOLOGY AND AGRICULTURAL GEOGRAPHY

An understanding of ecosystem analysis can be gained from:

BILLINGS, W. D. (1964) *Plants and the Ecosystem*, Macmillan, London.

ODUM, E. P. (1959) *Fundamentals of Ecology*, 2nd edn., Saunders, Philadelphia.

ODUM, E. P., (1963) *Ecology*, Holt, Rinehart, & Wilson, New York.

OVINGTON, J. D., HEITKAMP, D. and LAWRENCE, D. B. (1963) Plant biomass and productivity of prairie, savanna, oakwood, and maize field ecosystems in central Minnesota, *Ecology*, **44,** 52–63.

PHILLIPSON, J. (1966) *Ecological Energetics*, Arnold, London.

STODDART, D. R. (1965) Geography and the ecological approach: the ecosystem as a geographic principle and method, *Geography*, **50,** 242–51.

WESTLAKE, D. F. (1963) Comparisons of plant productivity, *Biological Reviews of the Cambridge Philosophical Society*, **38,** 385–425.

The interest of American anthropologists in "cultural ecology" is exemplified and discussed in:

CONKLIN, H. C. (1962) An ethnoecological approach to shifting agriculture, in WAGNER, P. L. and MIKESELL, M. W. (Eds.) *Readings in Cultural Geography*, U. of Chicago P., Chicago, pp. 457–64.

COWGILL, U. M. (1962) An agricultural study of the southern Maya lowlands, *American Anthropologist*, **64,** 273–86.

GEERTZ, C. (1963) *Agricultural Involution: the process of ecological change in Indonesia*, U. of California P., Berkeley and Los Angeles.

MIKESELL, M. W. (1967) Geographic perspectives in anthropology, *Annals of the Association of American Geographers*, **57,** 617–34.

STEWARD, J. H. (1955) *Theory of Culture Change*, U. of Illinois P., Urbana, pp. 30–42.

WOLF, E. R. (1966) *Peasants*, Prentice-Hall, Engelwood Cliffs.

An ecological approach to agriculture is either explicit or implicit in the work of the following geographers:

BLAUT, J. M. (1961) The ecology of tropical farming systems, *Revista Geografica*, **28,** 47–67.

GOUROU, P. (1966) *The Tropical World*, 4th edn., Longmans, London.

HARRIS, D. R. (1969) Agricultural systems, ecosystems and the origins of agriculture, in UCKO, P. J. and DIMBLEBY, G. W. (Eds.) *The Domestication and Exploitation of Plants and Animals*, Duckworth, London., pp. 3–15.

SAUER, C. O. (1956) The agency of man on earth, in THOMAS, W. L., JR. (Ed.) *Man's Role in Changing the Face of the Earth*, U. of Chicago P., Chicago, pp. 49–69.

SIMMONS, I. G. (1966) Ecology and land use, *Transactions of the Institute of British Geographers*, **38,** 59–72.

SPENCER, J. E. (1966) Shifting cultivation in southeastern Asia, *University of California Publications in Geography*, **19.**

CHAPTER 14

THE ECONOMIC
GEOGRAPHY OF AGRICULTURE

R. J. C. MUNTON

THE pace of methodological progress in agricultural geography has, until recently, been slow. Much of the literature has remained descriptive, similar in its conceptual framework to the regional studies of O. E. Baker and others in the 1920's and 1930's (see Chapter 13, above). Attempts at explaining land-use patterns have been couched primarily in terms of causal relations between agricultural activity and factors of the physical environment, with the result that agricultural geographers have largely ignored economic principles (Henshall, 1967; Coppock, 1968).

The scope of agricultural geography is restricted in this chapter to the production and sale of agricultural commodities. As a primary economic activity, agriculture may be investigated by means of economic principles. Alternatively, it can be examined ecologically by analysis of the biological and cultural relations within and between agricultural systems (see Chapter 13, above). Although descriptive studies have failed to provide a methodological framework for the evaluation of the influence of different aspects of rural life on agricultural production, they have brought non-economic considerations to the attention of geographers. For example, the notion that farming is a "way of life" has been contributed by the French school of human geography (Sorre, 1962). But, wherever agricultural production is oriented towards market demand, economic principles are required, both for the explanation of land-use patterns and for the objective assessment of non-economic influences.

THE CONCEPT OF ECONOMIC RENT

The concept of economic rent has been the basis of many agricultural location models designed to explain land-use patterns. The economic rent of a particular piece of land represents the net return yielded additional to that from land at the economic margin of farming. As such, it should not be confused with the rent paid by a tenant farmer. Both von Thünen and Ricardo established this concept in the 1820's, but their approaches were different. Ricardo recognized that variations in soil fertility would lead to variations in land use, resulting in differences in economic rent. Von Thünen, on the other hand, assumed soil fertility to be constant and the major variable cost to be the transportation of agricultural produce from farm to market. It followed that the type of land use and the intensity of cultivation were functions of distance from the market and, as a result, a series of concentric belts of differing land use were created around the market centre (Grotewald, 1959; Chisholm, 1962; Hall, 1968).

Von Thünen's emphasis on spatial relations has led geographers to develop his model, rather than that of Ricardo, and it therefore requires further consideration. Agricultural geographers neglected his work until the 1950's, partly because the emphasis he placed on transport costs seemed less justified in the twentieth century as they had become of reduced relative importance and, more particularly, some of his simplifying assumptions lacked reality. He envisaged, for example, an isolated state with a single, central market and a uniform transport network operating on a plain covered by soils of similar fertility. But the model still deserves our attention for at least two reasons. First, von Thünen's theory can explain contemporary land-use patterns around villages and farms in many parts of the world, although at this scale it is the labour cost, incurred by the time spent travelling from farmhouse to field, that is most significant (Chisholm, 1962). Second, the model represents an attempt to derive a theoretical explanation of land-use patterns, and, although qualified by von Thünen himself and later workers (Dunn, 1954; Garrison and Marble, 1957), his fundamental distance–cost relationship remains unchanged.

THEORETICAL CONCEPTS OF LAND USE

Certain implicit assumptions in von Thünen's land-use model cannot be fully justified and these have led to the development of more sophisticated concepts (Harvey, 1966; Henshall, 1967). For example, von Thünen's model made no provision for changes through time, a feature which can only be justified if all branches of technology advance at a uniform rate, thus causing the relative significance of individual costs to remain the same. In reality this is not so, as the relative decline in the importance of transport costs demonstrates. Furthermore, the spatial mobility of labour and capital result in continuous change in the relative significance of the factors of production (land, labour, capital) in any location and this leads to a continuous modification of the land-use pattern. Therefore, although static models have been used successfully in the planning of optimum regional locations for grain production in the United States (Egbert *et al.*, 1964), a dynamic approach is potentially more useful.

Garrison and Marble (1957) rejected the study of land-use regions because each region represents an amalgam of decisions taken by individual farmers with respect to their own farm circumstances; they therefore discuss location theory only within the farm context. Their most significant conclusion was that, in order to increase our understanding of land-use patterns, social costs must be integrated into our economic models. All previous work in agricultural location theory "has treated the entrepreneur as a person interested only in maximizing his net profits. Actually, experience has shown that maximization is more likely to be in terms of personal *satisfaction* than personal profit" (Garrison and Marble, 1957, p. 144). As a result, it has been argued that the "satisficer" concept—the idea that farmers take decisions that are "good enough"—is more appropriate (Harvey, 1966). Furthermore, empirical analysis suggests that not only do many farmers consciously make no attempt to maximize their income but, owing to a lack of ability and knowledge, most are incapable of taking the optimum decision from an almost infinite number of alternatives.

In their attempts to ascertain the reasons for the sub-optimum use of resources, Wolpert (1964) and Pred (1967) have also considered farmers' aims in life. These aims, for example, the desire to maximize profits, may well determine whether farmers adopt a new practice (Bowden, 1965) and thus they may significantly affect the rate of diffusion of an innovation within the rural community. For instance, Wolpert (1964) demonstrated that throughout central Sweden farmers are not maximizing their profits, partly because new ideas are adopted more slowly than changes are occurring in economics and technology.

In the light of these theoretical considerations, it is clear that the farm provides a most valuable focus for the economic geography of agriculture. There follows, therefore, a review of certain farm studies carried out by geographers and, as an example, a brief analysis is presented of the factors of production in relation to the British farm.

THE ANALYSIS OF FARMS

Chisholm (1964) has noted that the geographical study of farms has suffered from several limitations. It is impossible to employ aggregate land-use statistics to more than a limited extent in the explanation of land-use patterns, as these statistics cannot be related to the individual farm. The farm's land-use pattern, the basis of many geographical studies, represents merely the end result of a decision process, and in itself cannot provide evidence of the farmer's reasons for selecting particular activities (see Fig. 12).

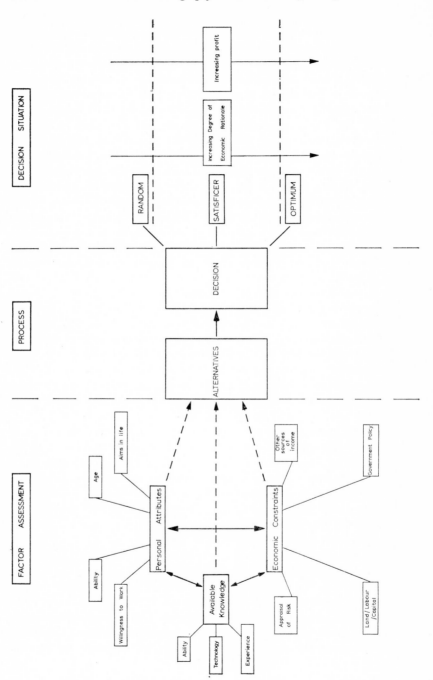

Fig. 12. Decision making on the farm.

Such information is best collected by personal interview of the farmer. Furthermore, most farm studies by geographers have been primarily concerned with the problems of classification and mapping and only secondarily with the analysis of the resulting spatial patterns. Too often attention has been concentrated on the classification of farm economies rather than on investigation of the interrelationships between aspects of the farm system.

Farms can be viewed as both ecosystems (see Chapters 4 and 13) and economic systems. In the ecosystem it is the flow of energy that is investigated; in the economic system it is the movement of money. Figure 13 shows diagrammatically the complex nature of the two systems and indicates that it is through the farmer (or rather his management decisions) that the two are integrated. There have been attempts to simplify models such as that shown in Fig. 13, especially through the statistical analysis of farm data (Heller, 1964; Henshall and King, 1966). But unless the correct data are collected, attempts at simplification can be misleading. For example, the analysis of data from individual fields can produce invalid results if it is done outside the context of the farms to which the fields belong, for, as integral parts of a larger business unit, decisions taken concerning their use may be related to overall farm policy.

The economic significance of production factors cannot be evaluated without an understanding of the kinds of management decision taken by the farmer. Major planning decisions, such as those concerning the balance between crops and livestock and the deployment of capital, are taken well in advance of their implementation. These decisions combine to form the optimum programme a farmer can, or is willing to, devise for his farm. Unpredictable adverse circumstances, such as staff illness and unfavourable weather, often result in detailed alterations. The plan and its modifications are implemented by organizational decisions taken on a weekly or daily basis. Both planning and organizational decisions may be reflected in the farm's enterprise structure, and because the farmer's policy is constantly being adjusted, the analysis of a single season's land use may be misleading. Furthermore, crops within a crop rotation are often related to each other. In Britain, for example, the selection, management, and yield of a crop are often more closely related to the previous crop than to any individual characteristic of the physical environment.

LAND, LABOUR, AND CAPITAL

The physical characteristics of land have often been treated as primary causes of land-use patterns. It has been suggested that maps of land use and particular aspects of the physical environment should be analysed first, and only if it was then felt necessary should human phenomena be studied

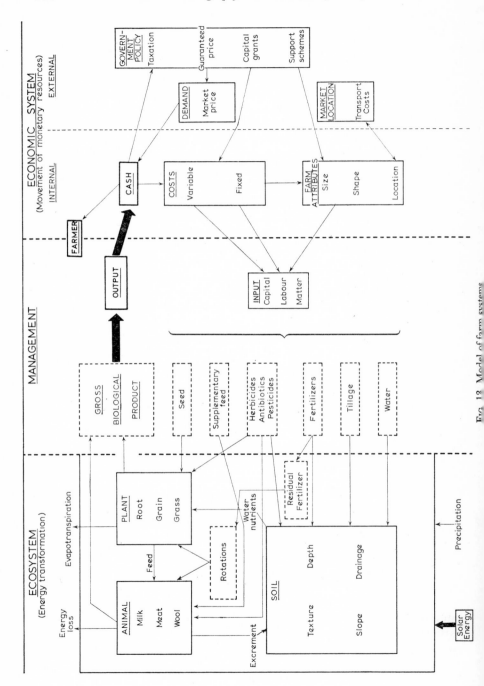

Fig. 12. Model of farm systems

(Buchanan, 1959). This approach cannot be justified for a number of reasons. First, land is only one factor taken into account by the farmer (see Fig. 12) and, as a result, may have little bearing on the decisions he takes. Second, its study in isolation from the farm makes an accurate calculation of its economic significance impossible. Third, the separate consideration of individual physical characteristics ignores their interdependence and in so doing may lead to an unrealistic evaluation of the total environment confronting the farmer. Fourth, each farmer may make a unique appraisal of similar land characteristics (Burton, 1962; Saarinen, 1966). Thus, although land has numerous measurable physical characteristics, attention should be focused on those that are interpreted by farmers as affecting their farm organizations. In other words, if the agricultural significance of land is to be meaningfully evaluated, it cannot be examined outside the economic, social, and technological contexts in which it belongs.

With improved technology, the physical and biological characteristics of land now present fewer constraints than formerly on the nature and intensity of land use, but certain factors, such as shallow soils, are still very limiting. Exceptions to this generalization suggest that the desire to crop more intensively may be outstripping our technological ability to do so. A well-publicized exception is the widespread presence of potato-root eelworm (*Heterochera rostochiensis*) in the English fen peat soils, which has enforced the adoption of a more extensive cropping system based on a reduced acreage of potatoes. Though land now presents fewer physical obstacles to farming, it is not losing its economic significance. Rising land prices necessitate greater capital investment and, as a result, the direct and indirect costs of land are increasing as a proportion of total costs.

The areal distribution of farm sizes in England and Wales has recently been described (Grigg, 1963), but little attempt has been made to establish the effect of farm size on land-use patterns. Unless there is an outside source of income the small farmer needs to employ enterprises that realize high net profits per acre, such as vegetables, potatoes, and dairying, in order to obtain a satisfactory living. Also, the relationship between land use and the physical environment may be affected by farm size. Rising costs have made economies of scale particularly important and it could be predicted that adjustments of land use to the physical environment should be more in evidence on large farms where economies of scale can be achieved in several enterprises.

Investigations of the relative importance of labour costs in the cost structure of farms, and the effect of labour costs on land use, have not been carried out by geographers, although Coppock (1965) has estimated regional agricultural labour requirements from land-use statistics. The effect of a declining labour force has been to increase the rate at which machinery is being substituted for labour, and such labour-intensive enterprises as dairying, potatoes, and sugar beet, have become concentrated on fewer farms, as farmers seek the necessary economies of scale to justify capital expenditure on machinery and the employment of expensive, skilled labour.

A more business-like approach to agriculture in recent years has resulted in farmers appreciating more fully the implications of the considerable, and increasing capital requirements of farming. Attempts to overcome falling returns on fixed-capital investment are discernible through the intensification of land use by, for example, heavier stocking of grass and continuous cereal cropping. More intensive land use, however, requires more sophisticated techniques and greater investment. Indeed, the long-term consequence of a continued fall in return on investment will be a depressed industry and fundamental changes in the land-use pattern.

CONCLUSION

Recent literature indicates that a reassessment of accepted concepts in agricultural economic geography is required. Greater examination of the economic significance of production factors is necessary if spatial variations in the land-use pattern are to be explained, particularly as non-economic motives are difficult to ascertain and impossible to measure, except as deviations from an expected economic pattern. Although farm visits are time-consuming, farms are the most obvious source of data, but their analysis, as complex systems, is likely to rely increasingly on multivariate statistical techniques and computers.

It has long been taken for granted that the study of land use is a relatively simple task, but there is no foundation for such an assumption. Findings from economics, biology, sociology, and psychology are increasingly being employed to determine the correct bases for our concepts of agricultural location. It is around these concepts that an economic geography of agriculture must be built, since without valid theories of agricultural production it is impossible to reduce the innumerable details of farming practice to some kind of order.

REFERENCES AND FURTHER READING

BACKGROUND LITERATURE

The following relate specifically to agricultural geography:

Coppock, J. T. (1964) Postwar studies in the geography of British agriculture, *Geographical Review*, **54**, 409–26.

Coppock, J. T. (1964) *An Agricultural Atlas of England and Wales*, Faber & Faber, London.

Coppock, J. T. (1968) The geography of agriculture, *Journal of Agricultural Economics*, **19**, 153–69.

Courtenay, P. P. (1965) *Plantation Agriculture*, Bell, London.

Henshall, J. D. (1967) Models of agricultural activity, in Chorley, R. J. and Haggett, P. (Eds.) *Models in Geography*, Methuen, London, pp. 425–60.

REEDS, L. G. (1964) Agricultural geography: progress and prospects, *Canadian Geographer,* **8,** 51–64.

SYMONS, L. (1967) *Agricultural Geography,* Bell, London.

More general texts by both geographers and agricultural economists include:

BROOKFIELD, H. C. (1964) Questions on the human frontiers of geography, *Economic Geography,* **40,** 283–303.

CHISHOLM, M. (1966) *Geography and Economics,* Bell, London.

CLARK, C. and HASWELL, M. R. (1964) *The Economics of Subsistence Agriculture,* Macmillan, London.

CLAYTON, E. S. (1964) *Agrarian Development in Peasant Economies: some lessons from Kenya,* Pergamon, Oxford.

DUCKHAM, A. N. (1963) *Agricultural Synthesis: the farming year,* Chatto & Windus, London.

DUMONT, R. (1957) *Types of Rural Economy: studies in world agriculture,* Methuen, London.

SORRE, M. (1962) The concept of *genre de vie,* in WAGNER, P. L. and MIKESELL, M. W. (Eds.) *Readings in Cultural Geography,* U. of Chicago P., Chicago, pp. 399–415.

Examples of early, descriptive studies are given in the bibliography of Chapter 13.

THE CONCEPT OF ECONOMIC RENT

CHISHOLM, M. (1961) Agricultural production, location and rent, *Oxford Economic Papers,* **13,** 342–59.

CHISHOLM, M. (1962) *Rural Settlement and Land Use,* Hutchinson, London.

DUNN, E. S. (1954) *The Location of Agricultural Production,* U. of Florida P., Gainesville.

GARRISON, W. L. and MARBLE, D. F. (1957) The spatial structure of agricultural activities, *Annals of the Association of American Geographers,* **47,** 137–44.

GROTEWALD, A. (1959) Von Thünen in retrospect, *Economic Geography,* **35,** 346–55.

HALL, P. (Ed.) (1968) *Von Thünen's Isolated State,* an English edition of *Der Isolierte Staat,* 1826, by J. H. von Thünen, translated by C. M. Wartenberg, Pergamon, Oxford.

THEORETICAL CONCEPTS OF LAND USE

A review of development in this field is given in:

HARVEY, D. W. (1966) Theoretical concepts and analysis of agricultural land-use patterns in geography, *Annals of the Association of American Geographers,* **56,** 361–74.

More specific aspects are illustrated by:

BOWDEN, L. W. (1965) Diffusion of the decision to irrigate, *University of Chicago Geography Research Paper,* **97.**

EGBERT, A. C., HEADY, E. O. and BROKKEN, R. F. (1964) Regional changes in grain production, *Iowa Agricultural and Home Economics Experiment Station Research Bulletin,* **521.**

PRED, A. (1967) Behaviour and location (part 1), *Lund Studies in Geography,* series B, **27.**

WILLIAMS, W. M. (1963) The social study of family farming, *Geographical Journal,* **129,** 63–75.

WOLPERT, J. (1964) The decision process in a spatial context, *Annals of the Association of American Geographers,* **54,** 537–58.

THE ANALYSIS OF FARMS

An account of the theoretical problems of farm-type classification is presented in:

CHISHOLM, M. (1964) Problems in the classification and use of farming-type regions, *Transactions of the Institute of British Geographers,* **35,** 91–103.

Two examples of farm classifications are:

BIRCH, J. W. (1964) Observations on the delimitation of farming-type regions, with special reference to the Isle of Man, *Transactions of the Institute of British Geographers*, **20**, 141–58.
JACKSON, B., BARNARD, C. and STURROCK, F. (1963) *The Pattern of Farming in the Eastern Counties*, Farm Economics Branch, University of Cambridge, Cambridge U.P., Cambridge.

Farms have been analysed in the following papers:

BLAUT, J. M. (1953) The economic geography of a one-acre farm in Singapore: a study in applied micro-geography, *Malayan Journal of Tropical Geography*, **1**, 37–48.
HELLER, C. F. (1964) The use of model farms in agricultural geography, *Professional Geographer*, **16**, 20–3.
HENSHALL, J. D. and KING, L. J. (1966) Some structural characteristics of peasant agriculture in Barbados, *Economic Geography*, **42**, 74–84.

A short, readable text on farm-management problems is that by:

DEXTER, K. and BARBER, D. (1968) *Farming for Profits*, 2nd edn., Iliffe, London.

Two standard references employing land-use data are:

COPPOCK, J. T. (1964) Crop, livestock and enterprise combinations in England and Wales, *Economic Geography*, **40**, 65–81.
STAMP, L. D. (1962) *The Land of Britain: its use and misuse*, 3rd edn., Longmans, London.

LAND, LABOUR, AND CAPITAL

There is no geographical analysis of these factors in relation to land-use patterns. A standard but heavy text by an agricultural economist is:

HEADY, E. O. (1952) *Economics of Agricultural Production and Resource Use*, Prentice-Hall, New York.

Geographical literature relating to this section includes:

BUCHANAN, R. O. (1959) Some reflections on agricultural geography, *Geography*, **44**, 1–13.
BURTON, I. (1962) Types of agricultural occupance of flood plains in the United States, *University of Chicago Geography Research Paper*, **75**.
COPPOCK, J. T. (1965) Regional differences in labour requirements in England and Wales, *Farm Economist*, **10**, 386–90.
GRIGG, D. B. (1963) Small and large farms in England and Wales: their size and distribution, *Geography*, **48**, 268–79.
SAARINEN, T. F. (1966) Perception of the drought hazard on the Great Plains, *University of Chicago Geography Research Paper*, **106**.

Attempts to relate land use to the physical environment include:

STEVENS, A. J. (1959) Surfaces, soils and land use in north-east Hampshire, *Transactions of the Institute of British Geographers*, **26**, 51–66.
TAYLOR, J. A. (1952) The relation of crop distributions to the drift pattern in south-west Lancashire, *Transactions of the Institute of British Geographers*, **18**, 77–91.

CHAPTER 15

NEW RESOURCE EVALUATIONS

G. MANNERS

RESOURCE EVALUATION AND THE GEOGRAPHER

A burgeoning world population, together with the quickening pace of economic and technological development, are making increasing demands upon the natural resources available to man. Although a serious long-term shortage of the major metals or conventional fossil fuels has yet to occur, the persistent growth in global requirements for such materials as iron ore, bauxite copper, tin, oil, and natural gas—plus occasional local scarcities—inevitably raise questions concerning their future adequacy. It is a matter of such fundamental importance at all stages and scales of economic development, that the thorough and recurrent investigation of resource availabilities has become an economic and political imperative.

Geographers have perennially laid great stress upon the importance and the quality of nature's endowments, and of man's selective—if at times indiscriminate—use of them. Consequently one might have expected that a lively discussion of resource availabilities would figure prominently in their literature. Yet evaluative judgements concerning the size and the quality of natural resources in general, and of mineral and energy resources in particular, are remarkably elusive in geographical writings. The received opinions of geologists and mining engineers concerning such endowments have been readily embraced, but the perceptive interpretation of their partial conclusions in the light of the spatial, economic, and other constraints upon resource use has been conspicuously absent. These constraints must be accorded full recognition in any complete appraisal of a resource, for without them a discussion of its size and worth can lead to a gross misinterpretation of its significance. Simpson (1966) affords a notable exception; but the somewhat perfunctory consideration given to the highly competitive market within which the British coal industry has to operate has occasioned a vivid contrast between the waning fortunes of the National Coal Board, and the infinitely more optimistic view taken of the coal industry's prospects (and hence the size and meaning of its resource base) in some recent geographical studies (Beaver, 1964; Humphrys, 1964; Thomas, 1961).

These and similar misjudgements originate fundamentally from a continuing failure on the part of many writers to distinguish clearly between "resources" on the one hand, and "geological occurrences" or "mineral deposits" on the other. What exactly do we mean when we introduce "an oilfield" into a discussion, or place a "hydro-electricity site" upon a map? The variable quality of resources is a commonplace. Without doubt, there is a general awareness that, irrespective of the tonnage of coal actually mined, the coalfields of Britain are not the resources they once were; there are even some who might argue that, far from being a resource, many of them are today a national liability. Probably, too, the natural gas deposits of the North Sea are widely interpreted as a much more valuable resource than their counterparts in North Africa or the Middle East. School and more advanced geography is frequently taught with these differences in mind. But is it not possible to specify the variations in the worth of particular resource endowments with rather greater precision than in the past? Are there not styles of thinking, research, and teaching which might permit an increasingly perceptive interpretation of resource availabilities?

It is certainly arguable that the reluctance of geographers to enter into those matters of public debate which touch upon natural resources suggests some deficiency in the classical tools of their trade. Comparing their historical commitment to matters concerning the environment and natural resources with (for example) their limited—albeit distinguished—contribution to the North Sea gas debate (Odell, 1966; Odell and Thackeray, 1967), or indeed to the whole question of national energy policy, the value of the geographer's traditional approach towards these matters must certainly be questioned. Only with the adoption of more analytical modes of inquiry will students and teachers of the subject be able to enter more fully into the increasing number of public discussions which are either related to, or have at their core, natural resource issues.

THE COMPLEXITY OF RESOURCE APPRAISAL

The problem of resource appraisal can best be approached by considering a specific mineral, iron ore. On a world scale it is subject to rapidly increasing demands. Measured in terms of its iron content,* global requirements grew from 116 million tons in 1950 to about 330 million tons 15 years later. By 1980 world demands will have reached 560 million tons (Manners, forthcoming). Thus, over a 30-year period, the consumption of iron ore will have increased nearly five-fold at an annual average rate of 5·4 per cent. Simultaneously, the location of these demands has undergone, and will continue to

* This is the most satisfactory means of equating ores of varying quality; subsequently it is abbreviated to tons (*c*).

TABLE 3. THE DEMAND FOR IRON ORE, BY REGIONS, 1950–80
(mill. tons (*c.*))

	1950	1964	1980
Total	116	331	560
North America	55 (47%)	73 (24%)	100 (18%)
Western Europe	29 (26%)	74 (23%)	105 (19%)
Eastern Europe	25 (21%)	87 (28%)	175 (31%)
Asia	4 (3%)	42 (14%)	132 (24%)
Oceania	2	4	8
Latin America	1 (3%)	6 (5%)	24 (9%)
Africa and Middle East	1	3	16
Unallocated production	—	22 (7%)	—

SOURCE. United Nations, 1968; Manners, forthcoming.

undergo, radical changes. Although the North American need for iron ore continues to increase, its share of world demand is steadily falling, as is that— although to a lesser degree—of Western Europe. In contrast, the demands of Eastern Europe are expanding both relatively and absolutely, whilst the increase in Asian requirements has been particularly rapid and for several decades appears likely to remain so (Table 3). In seeking to determine whether the world's reserves of iron ore will be sufficient to meet these needs both globally and regionally in 1980, the first aspect of the problem of resource evaluation is met.

Questions Concerning the Size of Reserves

A survey conducted under the sponsorship of the United Nations and published in 1950 estimated that the world's iron ore reserves were 27,000 million tons (*c*); by 1955, a second United Nations' estimate raised the figure to 42,000 million tons (*c*); and within another 4 years, Percival (1959) increased the latter figure threefold to 132,000 million tons (*c*). Such increments to the known world reserves of iron ore are easily explained. Fears of forthcoming ore shortages in the 1940's led to new technologies—such as sintering and pelletizing—which permitted the economic use of previously unacceptable ores. They also encouraged improvements in the transport of ore to allow its competitive shipment over longer distances (Manners, 1967). Simultaneously, on behalf of their ferrous industries, American and European geologists embarked upon extensive searches for new iron ore deposits in Canada, Latin America, and Africa, and found huge untapped resources there. The governments of the Soviet Union, China, and India also explored for and discovered new ore bodies for their expanding iron and steel industries, and were eventually able to contemplate substantial exports. And many countries of the developing world, seeking to find and exploit new mineral resources in their search for economic growth, also discovered iron ores in considerable quantities. Hence the increments to world reserves in the 1950's.

It would have been reasonable to expect that the estimates of iron ore reserves would continue their upward trend even after 1959, since a host of new and large deposits have been discovered on every continent during the 1960's. Yet when the United Nations (1968) attempted its third post-war survey in 1966, global reserves were put at 114,000 million tons (c), considerably below the 1959 figure suggested by Percival. Such a reduction in the apparent availability of iron ore in a period when new discoveries and technological advances were adding lavishly to known ore resources, raises doubts upon whether those concerned with the appraisals at different times were agreed on a definition of what constitutes an iron ore reserve.

Variations in Quality and Value

Iron ores vary considerably from deposit to deposit. They vary in their iron content, from the high-grade magnetites with 68–9 per cent iron, to the low-grade siderites and laterites with less than 30 per cent iron. They also vary in their chemical composition and physical structure. Consequently, the worth of an ore to an iron and steel industry anxious to produce carefully controlled qualities of pig iron at the lowest possible cost varies also. The price structure of iron ore reflects this. In 1965–6, for example, the Western European coastal iron and steel industry was prepared to pay about $16 per ton c. & f.* for pellets of 64 per cent iron, whereas it valued iron ore fines with 60–2 per cent iron at only approximately half that price.

The worth of a particular ore deposit is influenced not only by the quality of the products which can be produced from it, but also by its size and location. Small deposits, no matter how attractive their physical and chemical properties, excite very little interest from a modern mining company, dependent upon exploiting scale economies in production. Large and remote deposits, such as the "Superior"-type ores of Bolivia which have been estimated at 11,000 million tons (c), are similarly of little value until they can be economically transported to a large market. In other words, identical ores in terms of iron content, physical characteristics, and chemical composition are given quite different evaluations by the iron ore and the iron and steel industries at their source. These evaluations vary with time. They vary with changes in the technology of both mining and iron and steelmaking. They vary with movements in transport costs. And they vary with shifts in the supply and demand for ores at their markets. A recognition of this

* Carriage and freight; i.e. delivered to the blast furnace.

differing worth of ore deposits must therefore be built into resource appraisals. It is no simple matter.

Recognizing the Indeterminates

Another challenge which is presented in resource evaluation stems from the inherent limitations to the accuracy with which reserves can be quantified. Geological, economic, technological, and political indeterminates demand full recognition. For example, the geological aspects of an appraisal must rest upon a series of plausible assumptions about the underground geology in the district containing a deposit. These assumptions are liable to be proved inaccurate following a fuller exploration. However skilfully a geologist performs his task in quantifying the size of an iron ore deposit, at best he can only produce a rough estimate qualified by many uncertainties. Certainly the margin of error inherent in preliminary geological investigations is too great for mining companies; they always insist upon mining tests and more positive data on an ore body before they commit themselves to investments in mining plant, transport facilities, and the like. Such tests are costly. Since they can only be justified by a fairly firm intention to initiate ore exploitation, geological estimates of the size of iron ore deposits must mostly be made without them and on the basis of preliminary surveys only.

Although existing market conditions provide a reasonably accurate initial bench mark, the economic indeterminates of resource appraisal present even greater difficulties to accurate quantification. Current market prices for particular types of ore, plus the existing pattern of freight rates, suggest a range of prices which an ore is likely to command at a particular mine. Hence, they indicate the maximum level of mining costs admissible if that mine is to show a profit. Only with detailed mining tests can reasonably full production cost estimates be produced. Consequently an economic appraisal is faced initially with much the same problems as its geological counterpart. In addition, however, since it is quite unrealistic to rest the economic evaluation of an ore body upon contemporary market conditions, forecasts must also be made concerning the probable changes in ore demands, the prospective development of supplies, the trend in market prices, and the most likely movements in freight rates. The precision of such forecasts is clearly limited, and with it the accuracy of resource appraisal from an economic viewpoint.

Technological indeterminates are closely related to those of geology and economics. New processes in the iron and steel industry cause revisions to be made in the worth of particular ores. The recent, world-wide growth of L.D. oxygen steelmaking* has increased the relative importance of low-phosphorus

---

* A basic oxygen process named after the Austrian steelworks at Linz and Donawitz where it was pioneered.

ore demands to the disadvantage of the high-phosphorus ore producers. Similarly, technological progress in ore beneficiation and transport can significantly alter the competitive position of certain deposits in particular markets. Only a decade ago, for example, suggestions that Minnesota taconites would be pelletized on a large scale for the Mid-West market of the United States, or that large quantities of Brazilian ore would be shipped the 21,000 km to Japan in 100,000 wt carriers, would have been sceptically received; yet both are happening today.

Nor can political uncertainties be evaded in resource appraisal. For example, Canberra's decision in 1960 to allow the export of iron ore from Western Australia has transformed resource evaluations in many parts of the world. The fall in ore prices which accompanied the development of those fields, and the fact that they will have pre-empted about one-third of the Japanese market in the early 1970's have considerably modified the outlook for Indian and Peruvian exports. Again, should both the Soviet Union and China elect to offer large quantities of low-priced ore in Western Europe and Japan in the 1970's, and should their offers be accepted, the worth of the iron ore resources of, say, Brazil and West Africa, would be adversely affected. Whether the Soviet Union and China will opt for such a trading strategy is uncertain and, fundamentally, a political matter—although it would not be unrelated to their needs for foreign exchange.

The Taxonomy of Resources

A further aspect of the complexity of resource appraisal arises directly out of the earlier comments. Ideally, any evaluation of resources must relate to a reasonably articulate set of market circumstances in time and place—circumstances moulded by an intricate set of economic, technological, and political factors—and must be qualified by the varying degrees of accuracy with which they can be measured. Resource evaluations have always incorporated a crude taxonom. The earliest ones mirrored the interests of the geologists and mining engineers; who stressed the indeterminate aspects of underground geology in their appraisals; whilst the other constraints and uncertainties were either under-estimated or ignored. The terms used—"proved", "probable", and "possible"—clearly reflected this background. As a result, the suggestion of Blondel and Lasky (1955) that the term "reserves" should be limited solely to those deposits which can be exploited under existing economic conditions represented a distinct conceptual advance, even though it still failed to underline the importance of parallel technological and political constraints in resource evaluation.

If iron ore reserves are defined as those deposits which are suitable for profitable exploitation under existing (economic, technological, and political) conditions, then other deposits which are capable of profitable exploitation only when costs are lowered or prices raised can be placed in a separate category of "potential reserves". Each of these two categories, however, demands a further subdivision. Within the reserves of ore, it is important to distinguish between those quantities which have been carefully "measured", and others which are "inferred" from rather less precise evidence. Similarly there is a case for distinguishing between those potential reserves which are "marginal"—and appear capable of profitable exploitation with a slight (10 per cent?) increase in price, or from a small (10 per cent?) drop in production costs, through the introduction of new technology or with a change in political imperatives—and those which are "sub-marginal". The distinction may lack precision, but it has to be recognized that the latter would only become potential reserves with a substantial change in market and other

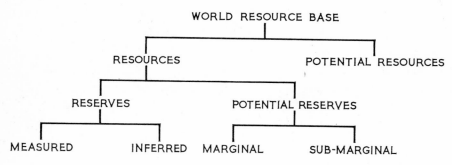

Fig. 14. The taxonomy of natural resources.

conditions. Further, it is not unreasonable to distinguish all these reserves from those "potential resources" which are the least attractive ore bodies remaining in the world's resource base (Fig. 14).

A SYSTEMS APPROACH TO RESOURCE EVALUATION

The principal conclusion to emerge from this examination of natural resource evaluation is not so much its taxonomy, its complexity, or the difficulties facing any attempt to quantify or qualify reserves. Rather it is to stress that natural resources can be studied realistically only as part of an intricate and dynamic system of interrelated phenomena which mould their demand, their supply, and their allocation. Technological, economic, transport, and political factors influence the size and the nature of resource requirements. Geological, technological, economic, transport, and political matters constrain their availability. And these two set of forces are reconciled through allocation processes in which either a market-guided price structure or, in the fully planned economies, government decisions, reconcile the needs of consumers with the possibilities of production (Fig. 15). For each resource occurrence, therefore, an evaluation of its "reserves" and "potential reserves" must rest upon a recognition of this changing system of demand influences, supply constraints, and market allocation processes.

The practicability of such an approach to resource appraisal depends largely upon the scale of review. At the world scale the task is usually not feasible— even if a static market framework is assumed—as a consequence of the enor-

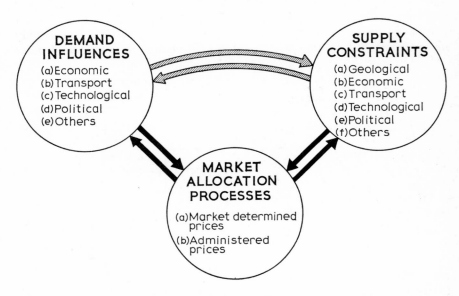

Fig. 15. Schematic summary of the resource system.

mous size and variety of most resources, and the survey costs involved. More generalized resource evaluations have to suffice at this scale. Fortunately, there appears to be little urgency for detailed global surveys of most key natural resources, since the evidence available suggests their overall adequacy. For example, the somewhat loosely measured iron ore "reserves" of the 1966 United Nations study suggest that at least 114,000 million tons (c) of ore appear to be readily available to the world iron and steel industry; with global demands for iron ore standing at about 530 million tons (c) in 1965, such a figure suggests the overall adequacy of this important resource in the foreseeable future. The capacity of existing mines and proposed mining developments together indicate that at least 640 million tons (c) of annual mining capacity will be available in 1980, whereas demand is likely to be only 560 million tons (*c*) (United Nations, 1968; Manners, forthcoming). Certain mines are assured outlets because of their ownership, their long-term contracts, or their rôle in a particular economy. Therefore the amount of mining capacity likely to be available throughout the 1970's to satisfy uncommitted demands will considerably exceed requirements. As a result some plans for mine expansion may be deferred; other mines may be closed; certainly the price of iron ore will continue to drift downwards. The general decline in the value of iron ore therefore means that the 1966 United Nations' evaluation of ore reserves strictly also needs some downward adjustment. Nevertheless, the global iron ore situation gives no cause for apprehension as far as the adequacy of reserves is concerned.

Nationally and regionally the situation is rather different. As the world's iron and steel industry becomes increasingly dependent upon basic oxygen steelmaking with its need for low-phosphorus ores, a re-evaluation of the importance and, indeed, the size of Swedish high-phosphorus resources becomes essential. Swedish ore exports are by no means about to collapse, of course. There remain many important centres of demand for high-phosphorus ores, and their persistence can probably be assured by Swedish exporters opting for attractive pricing policies which will encourage investments in L.D.–A.C.* steelmaking technology. It is also technically possible to reduce the phosphorus content of many Swedish ores through pelletization. However, the changing pattern of ore demands certainly means that the rate of growth of Swedish ore exports will be checked, that the increasing value of those exports will be even more restrained, and that a re-evaluation of the country's ore resources would not be out of place. Again, in Lorraine, a reassessment of one of the world's most famous ore deposits is now overdue. Despite its low production costs per ton, the lean grade of the Minette ores and the high costs of smelting them into pig iron throw doubts upon their medium-term value. The blast furnaces of eastern France are already being enriched with high-grade ores imported from overseas via the Rhine and newly canalized Moselle. The major remaining uncertainty is the speed at which imports will be completely substituted for local ores—and hence the exact date when the ores of the Lorraine basin become only "potential reserves".

Similarly, in Britain, difficulties are already confronting the lean ore producers of the East Midlands. As the British Steel Corporation is restructured geographically, it is highly probable that a diminishing share of national pig iron production will be based upon East Midland ores. How should these ore resources be assessed today? Should the Lincolnshire and Northamptonshire ore fields be drawn upon our maps as they have been in the past?

---

* A modified version of L.D. process which has slightly higher costs but allows the use of high-phosphorus ores.

Will there be a home ore industry in 10 years' time? Only a careful and systematic analysis of the existing and the prospective markets for the Jurassic ores, in the light of their prospective production costs and the intensification of overseas competition, will begin to provide an adequate answer. It is surely to such assessments that geographical studies should be turning.

This chapter has sought to illustrate the need for a more sophisticated interpretation of natural resource availabilities in general, and of mineral and energy resource distributions in particular. The plea for a "systems approach" represents more than the adoption of a contemporary phrase. It seeks to draw attention to the narrow and blinkered style of many resource evaluations in the past. Economists have made good conceptual progress in recent years towards the better evaluation of resources from their own distinctive viewpoint (Schurr and Netschert, 1960; Landsberg *et al.*, 1963). By adding a more articulate spatial component to these studies, geographers will be in a much stronger position not only to clarify their own interpretation of nature's endowments, but also to give a stronger lead in those debates on natural resources which increasingly command public attention.

REFERENCES AND FURTHER READING

RESOURCE EVALUATION AND THE GEOGRAPHER

Especially noteworthy amongst recent major resource appraisals of minerals and energy are:

ADELMAN, M. A. (1964) World oil outlook, in Clawson M. (Ed.) *Natural Resources and International Development*, Johns Hopkins, Baltimore.

BROOKS, D. B. (1966) *Low Grade and Non-conventional Sources of Manganese*, Johns Hopkins, Baltimore.

BRUBAKER, S. (1967) *Trends in the World Aluminum Industry*, Johns Hopkins, Baltimore.

FISHER, J. L. and POTTER, N. (1964) *World Prospects for Natural Resources*, Johns Hopkins, Baltimore.

LANDSBERG, H. H. (1964) *Natural Resources for U.S. Growth*, Johns Hopkins, Baltimore.

LANDSBERG, H. H., FISCHMANN, L. L. and FISHER, J. L. (1963) *Resources in America's Future, Patterns of Requirements and Availabilities, 1960–2000*, Johns Hopkins, Baltimore.

The issues of resource definition are discussed in:

NETSCHERT, B. C. and LANDSBERG, H. H. (1961) *The Future Supply of Major Metals*, Johns Hopkins, Baltimore.

SCHURR, S. H. and NETSCHERT, B. C. (1960) *Energy in the American Economy, 1850–1975*, Johns Hopkins, Baltimore.

Geographical contributions to the study of resources include:

BEAVER, S. H. (1964) Mineral resources and power, in WATSON, J. W. and SISSONS, J. B. (Eds.) *The British Isles, a systematic geography*, Nelson, London.

HUMPHRYS, G. (1964) The coal industry, in MANNERS, G. (Ed.) *South Wales in the Sixties*, Pergamon, Oxford.

MANNERS, G. (1964) *The Geography of Energy*, Hutchinson, London.

ODELL, P. R. (1963) *An Economic Geography of Oil*, Bell, London.

ODELL, P. R. (1966) A three point approach necessary to exploit sea gas, *The Times* (21 July).
ODELL, P. R. and THACKERAY, F., (1967) The price of North Sea gas, *The Times* (7 December).
SIMPSON, E. S. (1966) *Coal and the Power Industries in Post-war Britain*, Longmans, London.
THOMAS, T. M. (1961) *The Mineral Wealth of Wales and its Exploitation*, Nelson, Edinburgh.
THOMAS, T. M. (1966) The North Sea and its environs: future reservoir of fuel?, *Geographical Review*, **56,** 12–39.

THE COMPLEXITY OF RESOURCE APPRAISAL

Further aspects of iron ore resources can be explored in:

BLONDEL, F. and LASKY, S. G. (1955) Concepts of mineral reserves and resources; in United Nations, *Survey of World Iron Ore Resources, Occurrence, Appraisal and Use*, New York.
British Iron and Steel Federation, 1963, Structural change in world ore, *Steel Review*, **30,** 18–30.
The Economist, 1963, Lorraine under sentence (9 March).
JOHNSON, J. A. (1967) Developments in the Swedish iron ore industry, *Geography*, **52,** 420–2.
MANNERS, G. (forthcoming), *The Changing World Market for Iron Ore, 1950–1980*.
MANNERS, G. (1967) Transport costs, freight rates and the changing economic geography of iron ore, *Geography*, **52,** 260–79.
METAL BULLETIN (1965) *Iron ore* (special supplement).
PERCIVAL, F. G. (1951, revised 1959) *The World's Iron Ore Supplies*, British Iron and Steel Federation, London.
POCOCK, D. C. D. (1966) Britain's post-war iron ore industry, *Geography*, **51,** 52–5.
United Nations (1950) *World Iron Ore Resources and Their Utilization*, New York.
United Nations (1955) *Survey of World Iron Ore Resources, Occurrence, Appraisal and Use*, New York.
United Nations (1968) *World Market for Iron Ore*, New York.

Resource issues and public policy are debated in:

LANDSBERG, H. H. and SCHURR, S. H. (1968) *Energy in the United States; sources, uses and policy issues*, Random House, New York.
Ministry of Power (1965) *Fuel Policy* (Cmnd. 2789).
Ministry of Power (1967) *Fuel Policy* (Cmnd. 3438).
Political and Economic Planning, (1963) An energy policy for EEC?, *Planning*, **29.**
Political and Economic Planning, (1966) *A Fuel Policy for Britain*, P.E.P., London.
Resources for the Future (1968) *U.S. Energy Policies, an Agenda for Research*, Johns Hopkins, Baltimore.

CHAPTER 16

PROGRESS IN TRANSPORT GEOGRAPHY

MARION W. WARD

IN THIS short chapter emphasis is given to a particular approach which is current in transport geography—the treatment of transport systems as networks. One of the most important developments in this field has been the application of graph theory to the study of transport networks, in order to measure geometrically the manner in which places in a network are connected by transport links. A second interesting development discussed here is the examination of the rôle of transport in economic growth, using a generalized scheme to describe the processes by which transport facilities grow coupled with the measurement of the relationships between the provision of transport and features of economic development. In addition, the three major modes of transport—sea, land, and air—are considered, as each has been the subject of studies in which mathematical techniques and models have been used to probe more deeply than was previously possible into such concepts as hinterlands and tributary areas, port and network growth, and the effects of new highway development.

TRANSPORT NETWORKS
AND GRAPH THEORY: AN EXAMPLE

The use of graph theory (part of the branch of geometry known as topology) in transport geography was first introduced by Garrison (1960) and followed up by his student Kansky (1963); the method, and network analysis in general, now has wider significance in geography as a whole (Chorley and Haggett, forthcoming). To make any network amenable to analysis by graph theory, each place is represented by a point, or vertex (v), and each route or connection between places is represented by a link, or edge (e). The network illustrated here is one for the air-transport routes flown by Fokker Friendship aircraft in Papua and New Guinea, represented in Fig. 16 as a graph with 10 vertices and 12 edges. Interest in such a network centres around two questions: (1) how well connected is the network viewed as a whole and (2) how well connected is each individual point with every other point in the network? Graph theory provides several topological measures which help to answer these questions.

The degree of connectedness, or connectivity, of the whole network can be described by the cyclomatic number, μ, which measures how many fundamental circuits there are in the network. A graph with only the minimum number of links needed to join every point into the network would have no fundamental circuits. A network containing the maximum number of circuits possible would have many redundant routes over and above those required for relatively efficient connections between points.

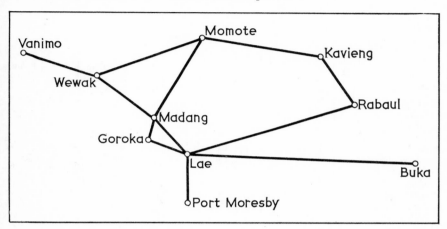

FIG. 16. Network of scheduled air services by Fokker Friendship aircraft, Papua and New Guinea, 1968.

In the example shown in Fig. 16:

$$\mu = e - v + p$$
$$= 12 - 10 + 1$$
$$= 3$$

where p = number of "networks", e = number of edges, v = number of vertices. That is to say there are three fundamental circuits (linking Lae, Rabaul, Kavieng, Momote, Wewak, Madang, and Goroka) and, since this number is considerably lower than the maximum possible number, the interpretation is that the connectivity of the system is relatively low.

The extent, or topological size, of the network is indicated by the diameter, δ, which is the smallest number of edges connecting the furthest pair of points. In this example the diameter = 4, and there are several possible locations for it (for example, from Vanimo to Rabaul or from Vanimo to Port Moresby).

The second question is concerned with the connectivity of individual points in a transport network. For example, which point in the Papua and New Guinea air network is most highly connected with all other points? One of the clearest measures of this is obtained by converting the graph into a connectivity matrix (or one-zero array) in which a 1 in a cell indicates that a connection exists between the row place and the column place, and a 0 indicates that no direct connection exists. Using matrix algebra multiplication, this matrix is raised to the power of the graph's diameter, yielding the desired information about each point (Fig. 17).

	Vanimo	Wewak	Momote	Kavieng	Rabaul	Madang	Goroka	Lae	PortMoresby	Buka	Row totals	Per cent	Rank
Vanimo	3	2	5	1	2	6	2	2	1	1	25	4·0	10
Wewak	2	14	9	7	3	11	8	12	2	2	70	11·2	4
Momote	5	9	15	3	7	12	9	11	3	3	77	12·3	3
Kavieng	1	7	3	7	1	8	3	9	1	1	41	6·5	7
Rabaul	2	3	7	1	9	10	8	3	6	6	55	8·8	6
Madang	6	11	12	8	10	25	12	13	8	8	113	18·0	1
Goroka	2	8	9	3	8	12	11	10	6	6	75	12·0	5
Lae	2	12	11	9	3	13	10	30	2	2	94	15·0	2
Port Moresby	1	2	3	1	6	8	6	2	5	5	39	6·2	8
Buka	1	2	3	1	6	8	6	2	5	5	39	6·2	8

FIG. 17. Total connectivity of points in the Papua–New Guinea air network.

The value in each cell (X_{ij}) represents the number of ways in which one may journey between the ith place and jth place in exactly four steps (because $\delta = 4$). If the rows are totalled and expressed as a percentage of the final column total, these percentage values give a measure of the total connectivity of each place. The best connected (or most accessible) place is Madang, followed by Lae. The least well connected of all the vertices is Vanimo, and the low values for Port Moresby and Buka also indicate poor connectivity.

There are available several measures like these for the analysis of networks treated as graphs and represented in matrix forms, but space does not permit their description (see Kansky, 1963; Chorley and Haggett, forthcoming). It it worth emphasizing that as the complexity of the network increases the value of using such measures becomes significantly greater. They provide perhaps the only precise means of analysing the total accessibility and linkage within the network, and although not yet widely applied, the use of graph theory represents a significant step forward in the analysis of transport patterns.

TRANSPORT AND ECONOMIC DEVELOPMENT

It has long been recognized that economic development is intimately related to the growth of efficient transport systems. Berry (1960), for example, used several simple measures of transportation as some of his criteria for the ranking of countries in terms of their economic development. A direct examination of the relationships between transport and economic progress was carried out by Taaffe *et al.* (1963) in West Africa.

The first part of their study postulated a general sequence of transport development which is here applied to Malaya over the last 100 years. The sequence commences with a scattering of small ports and trading posts along the seacoast. There is little lateral interconnection and each port has a limited hinterland (Fig. 18a). Major lines of inland penetration develop along railways from some ports to areas of economic growth (in this case), the tin fields and reduce hinterland transportation costs for these ports (Fig. 18b). As a result, the hinterlands of these major ports expand at the expense of the other ports, while interconnecting links begin to develop between interior centres and feeder services are established (Fig. 18c). The process of lateral interconnection may continue as economic growth goes on until all the ports, interior centres, and junctions on the transport network are linked (Fig. 18d). It is likely that while this is happening certain routes will have become dominant as "main streets", or corridors, and at this stage another cycle of transport development, associated perhaps with alternative modes of transport or higher quality services, may be superimposed on the first. In Malaya, other modes of travel subsequent to the railways are concentrated along such corridors, in the form of road and air services (Fig. 18e).

The second part of the Taaffe *et al.* study analysed the relationship in Nigeria and Ghana between transport provision and other characteristic features associated with different stages of economic development. By using regression analysis they showed that the greatest road densities are in the areas with most population—confirming a relationship which is fairly obvious, since the distribution of population itself reflects a number of other characteristics, such as land quality and resource distribution, to which road densities might well also be related. Five additional factors were found to be necessary for the full explanation of the spatial variation in road network density: hostile environment, rail competition, intermediate location, income, or degree of commercialization, and relationship to the idealized growth sequence already described.

Other measures of economic and transport development have also been

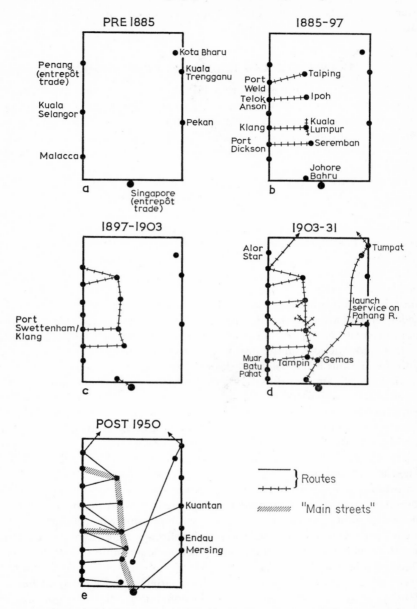

FIG. 18. The Taaffe–Morrill–Gould sequence of transport development applied to
Malaya.

used to study these relationships. For example, Yeates (1968) found a direct correlation between an index of railway connectivity for eighteen countries and their *per capita* incomes, and also between the size of population of a network vertex and the amount of traffic flowing through it.

These studies provide useful if limited explorations of the relations between transport evolution and economic growth, and attempt to quantify relationships which previously have been accepted on a more intuitive basis. The study by Taaffe *et al.* is particularly significant in pointing the way towards the construction of a model to simulate the economic growth process which will allow the evaluation of alternative transport development policies.

PROGRESS IN THE STUDY OF TRANSPORT MODES
Ports and Shipping

The student of port geography must still refer to Weigend (1958) for the definition of basic concepts such as those of port, foreland, hinterland, and maritime space, of which port and hinterland have received closest examination. A number of attempts at rational classification of ports have been made (Bird, 1963; Rimmer, 1966) and they have led to reconsiderations of the problems involved in measurement of port size and function. The study of port hinterlands has concentrated on definition and delimitation, providing interesting comparisons between actual and theoretically derived hinterlands and the examination of discrepancies between the two (Rimmer, 1967; Ward, 1966). This problem has great practical significance, illustrated, for example, by the rejection of the Port of Bristol Authority's plan for port expansion, a decision which was based at least in part on the results of a gravity model prediction of the port's future trade as it competes with other British ports (Ministry of Transport, 1966).

Interest in the process of port development led Bird (1963) to propound a descriptive scheme for the phases of growth of "Anyport", outlined in Table 4.

TABLE 4. PORT DEVELOPMENT SCHEME

Era	Terminated by
I Primitive	the overflowing of the port function from the primitive nucleus of the port or the change in location of the dominant port function
II Marginal quay extension	the change from a simple continuous line of quays
III Marginal quay elaboration	the opening of a dock or the expansion of the harbour
IV Dock elaboration	the opening of a dock with simple lineal quayage
V Simple lineal quayage	the provision of oil berths in deep water
VI Specialized quayage	the occupation of all waterside sites between the port nucleus and the open sea

SOURCE. Bird, J. (1963), *Seaports of the United Kingdom*, Hutchinson, London, p. 34.

Though this generalization is based on the estuarine ports of Western Europe, it can be readily adapted to fit the development of the natural deep-water harbours of the southern hemisphere or the major ports in colonial dependencies (Hoyle, 1968), and could well be linked with the outline of transport development from a coastal origin already discussed.

Road and Rail

The sequence of transport growth recognized by Taaffe *et al.* described four stages in what is really a continuous process, though there is some justification for distinguishing the major phases. Several attempts have been made to build mathematical models to simulate continuous network development in a more realistic fashion. Kansky (1963) and Morrill (1965) have tried to reproduce the location patterns of railway routes as they develop over time, dealing with small areas and simple networks in Sicily and central Sweden respectively. Morrill (1965) produced a simulated road pattern, developing from a central point, as part of a more complex model of the evolution of a central-place hierarchy. These and other attempts have met with mixed success, but they indicate an important direction in which progress is being made as our understanding of the processes of network growth and change improves.

Road and rail facilities often compete with each other for a share of the transport market, and Kanaan (1966) has examined the road and rail networks of Syria from this point of view. The connectivity of each network has changed through time, and Kanaan has related this to varying rates of economic development in different parts of the country, and has shown how some inland urban centres have declined in connectivity and others, such as new oil-handling ports, have increased.

Rapid expansion in car ownership and increased rates of highway construction in the United States and Western Europe have given rise in the last 15 years to intense interest in the economic and social consequences of these changes. Garrison *et al.* (1959) made an important early contribution to this work by showing how the impact of route changes on the urban hierarchy and on the location of business and service establishments could be evaluated. The effects of different levels of transport provision and of changes in travel habits are topics which require separate treatment. Sufficient to say here that the reader will encounter discussions of these matters in almost every branch of human geography, ranging from studies of employment and the journey to work to theories of urban growth, from the analysis of location patterns in retailing to problems in urban and regional planning.

Air Transport

This particular mode of travel has received relatively less investigation from geographers than other modes of transport, but several studies do exist which concentrate on the structure of the route network and on the delimitation of effective hinterland areas. Garrison and Marble (1962) used the graph-theory methods previously described to study Venezuelan domestic flight patterns. The connectivity matrix was large, and was analysed by the statistical method of components analysis in order to reduce the complexity of the network. This showed that the basic structure of the air network was dependent on: (1) the size hierarchy of cities in the network; (2) the metropolitan dominance of Caracas; (3) the existence of two major sub-regions; (4) the presence of a weak minor regionalization effect. The results of this and other studies tend at present to confirm patterns which can be recognized intuitively, but do so in a way which lends precision to our understanding and objectivity to our conclusions about the structure of route networks. In air transport the hinterland concept is also an important one, and Taaffe (1959 and 1962) and Perry (1968; see Chapter 19 below) have laid emphasis on this aspect for several major centres.

CONCLUSION

Future work in transport geography seems likely to explore further the use of techniques derived from graph theory to increase understanding of the spatial aspects of transportation networks. Geographers will also continue to be concerned with the spatial consequences of such technological innovations as supertankers, containerization, the growth of mass air transport, and the further development of motorways. It seems likely that greatest progress will be made in those fields most closely linked with the practical business of operating transport concerns, since it is here that new technical advantages pose problems which urgently demand solutions.

REFERENCES AND FURTHER READING

TRANSPORT NETWORKS AND GRAPH THEORY

CHORLEY, R. J. and HAGGETT, P. (forthcoming) *Network Models in Geography*, Arnold, London.

GARRISON, W. L. (1960) Connectivity of the Interstate Highway System, *Regional Science Association, Papers and Proceedings*, **6,** 121–37.

HAGGETT, P. (1967) Network models in geography, in *Models in Geography*, Methuen, London, pp. 609–68.

KANSKY, K. J. (1963) Structure of transportation networks; relationships between network geometry and regional characteristics, *University of Chicago Geography Research Paper*, **84.**

PITTS, F. R. (1965) A graph-theoretic approach to historical geography, *Professional Geographer*, **17,** 15–20.

YEATES, M. H. (1968) *An Introduction to Quantitative Analysis in Economic Geography*, McGraw-Hill, New York.

TRANSPORT AND ECONOMIC DEVELOPMENT

BERRY; B. J. L. (1960) An inductive approach to the regionalization of economic development, *University of Chicago Geography Research Paper*, **62**, 78–107.
GOULD, P. R. (1960) Transportation in Ghana, 1910–1959, *Studies in Geography*, **5**, Northwestern U.P., Evanston.
TAAFFE, E. J., MORRILL, R. L. and GOULD, P. R. (1963) Transport expansion in underdeveloped countries: a comparative analysis, *Geographical Review*, **53**, 503–29.
WILSON, G. W., BERGMANN, B. R., HIRSCH, L. V. and KLEIN, M. S. (1966) *The Impact of Highway Investment on Development*, The Brookings Institution, Washington.

PROGRESS IN THE STUDY OF TRANSPORT MODES

PORTS AND SHIPPING

BIRD, J. (1963) *The Major Seaports of the United Kingdom*, Hutchinson, London.
HOYLE, B. S. (1968) East African seaports: an application of the concept of "Anyport", *Transactions of the Institute of British Geographers*, **44**, 163–83.
Ministry of Transport (1966) *Portbury: Reasons for the Minister's Decision not to Authorise the Construction of a New Dock at Portbury, Bristol*, H.M.S.O., London.
RIMMER, P. J. (1966) The problems of comparing and classifying seaports, *Professional Geographer*, **18**, 83–91.
RIMMER, P. J. (1967) Inferred hinterlands: the example of New Zealand, *Geography*, **52**, 384–92.
WARD, M. W. (1966) Major port hinterlands in Malaya, *Tijdschrift voor Economische en Sociale Geografie*, **57**, 242–51.
WEIGEND, G. G. (1958) Some elements in the study of port geography, *Geographical Review*, **47**, 185–200.

ROAD AND RAIL

GARRISON, W. L., BERRY, B. J. L., MARBLE, D. F., NYSTUEN, J. D. and MORRILL, R. L. (1959) *Studies of Highway Development and Geographic Change*, U. of Washington P., Seattle.
KANAAN, N. J. (1966) Structure and requirements of the transport network of Syria, *Highway Research Record*, **115**, 19–28.
KANSKY, K. J. (1963) Structure of transportation networks, *University of Chicago Geography Research Paper*, **84**.
MORRILL, R. L. (1965) Migration and the growth of urban settlement, *Lund Studies in Geography, series B*, **26**.
STOKES, C. J. (1967) The freight transport system of Colombia, *Economic Geography*, **43**, 71–90.

AIR TRANSPORT

GARRISON, W. L. and MARBLE, D. F. (1962) The structure of transportation networks, *U.S. Army Transportation Command, Technical Report*, **62–11**.
PERRY, N. (1968) The air traffic of Jersey (Channel Islands), *Tijdschrift voor Economische en Sociale Geografie*, **59**, 156–64.
TAAFFE, E. J. (1959) Trends in airline passenger traffic: a geographic case study, *Annals of the Association of American Geographers*, **49**, 393–408.
TAAFFE, E. J. (1962) The urban hierarchy: an air passenger definition, *Economic Geography*, **38**, 1–14.

CHAPTER 17

PROGRESS IN INDUSTRIAL GEOGRAPHY

J. SALT

INDUSTRIAL geography attempts to describe and explain the locational patterns of manufacturing activity. Until recently the study has been more concerned with description than with explanation. What explanation there has been has concentrated on the relationship between industrial activities and the physical environment rather than economic and social phenomena. In fact, it is only since the mid-1950's that economic geography has incorporated much economic thinking. This change has been brought about by geographers and economists, both concerned with questions of location. Not only has the importance of spatial considerations been realized in such problems of economic development as regional income differences, but refinements of methodology have made it possible for these to be incorporated in the analysis of economic patterns.

The need for a theoretical basis to the study of industrial location has only recently been recognized. Industrial geographers have long been concerned with case studies of industries and regions but the subject has lacked the necessary conceptual framework to tie together the diverse strands of industrial activity, enabling its total spatial pattern to be viewed as a whole. "The economic geographer needs theoretical analysis to clarify his mind, to define his problem and to strip it of irrelevancies" (Smith, 1955, p. 1). Given this need it was natural that geographers should turn to the macro-analysis of economic theory to provide the conceptual framework in which to place case studies. From the marriage of economic theory with empirical geographical research into the factors affecting location and the peculiarities of particular industries and regions, principles of industrial location have been formulated.

LOCATION THEORY AND
THE CONCEPT OF OPTIMUM LOCATION

Within the body of economic theory used by industrial geographers, the concept of optimum location has been paramount. The idea that for all production units there should be a best possible location stems from the writings of such early economists as Smith (1776), Ricardo (1817), and Mill (1848), who assumed the existence of conditions of perfect or near-perfect competition. From such ideas, formal location theories were developed which attempted to balance the supply of production factors and the demand for products within an established areal pattern of production, that is to provide for a spatial equilibrium of activities. This equilibrium never actually exists, for price, a function of the supply–demand relationship, usually cannot adjust itself to one change in supply or demand before another occurs. Nevertheless, the concept of optimum location with spatial equilibrium is useful for it provides a standard against which to measure existing patterns.

Classical location theory, which determines the optimum location by minimizing transport costs, has developed from this basic concept. Deviations from the optimum can be explained in terms of variations in costs other than those of transport. The pioneering work of Weber (1909) was concerned with the principles by which costs could be minimized in the choice of location. His ideas have become known as the "least cost" or "cost minimization" theory. The optimum location is defined as the point at which transport costs per unit of production are least, or where the benefits of agglomeration and/or labour availability are sufficient to offset the disadvantages caused by not locating where the transport costs are least. Different industries will, of course, have different optimum locations depending upon the nature and volume of their inputs and outputs, and hence of their cost structures.

The main criticism of Weber's theory has been its primary concern with supply and its depiction of the market as existing at a particular point. Location is determined by analysing the pattern of inputs and does not take enough consideration of the strength and pattern of market areas. The main virtue of the theory lies in its recognition of the existence of transport and non-transport factors, such as agglomeration and labour, and of localized and ubiquitous materials. The location cost of a factor is that incurred in bringing it to the place of production. Ubiquitous materials, by their very nature, have no location cost and will not influence location.

Lösch (1940) attempted to incorporate demand into location theory by considering the size of market area. In his view the optimum location is that which commands the largest market area and thus brings in the highest possible revenue: his theory is one of "profit maximization". As Weber underemphasized demand, so Lösch tended to neglect supply. Subsequent workers like Isard (1956) and Greenhut (1956) have modified Lösch's ideas by attempting to embrace cost considerations more fully. In the resulting "minimax" solution, the optimum location is the point which combines lowest possible costs with highest possible revenue. Acceptance of spatial variations in both demand and supply makes the formulation of any overall theory of plant location, as attempted by Isard (1956), very difficult. But the use of such advanced mathematical techniques as linear programming in the determination of patterns is increasing our understanding of these variations, although an accurate general system of location is not yet available.

THE SIGNIFICANCE OF COST AND PRICE

The relationship between cost and price is fundamental to the study of industrial location for their variations set limits within which profitable production can be undertaken. Within these limits the individual entrepreneur is free to locate anywhere, although the size of his profits will vary according to the pertinent cost–price relationship. Smith (1966) used Weber's ideas to build a theoretical model for geographical studies of industrial location. It presents the cost–price situation in an industry in its simplest form and is a useful aid in the understanding of how a location decision is made. The basis of Smith's model is that total costs vary from place to place in accordance with variations in production factor and marketing costs. Demand and price also vary between areas and hence the total return on investment varies. In the best location, return will exceed costs by the greatest amount.

Figure 19 shows a simple, spatial cost–price situation. It assumes that both cost and price are fixed and cannot be altered by the individual firm, that output is constant in space, and that any variations in demand are reflected in areal price variations. Cost and price are plotted on the vertical axis, distance on the horizontal axis. In Fig. 19a costs are assumed to be variable and the average cost per unit of production at any point is represented by the cost line. Demand is assumed to be constant and therefore the price is the same everywhere. In this cost–price situation, 0 becomes the optimum location while *Ma* and *Mb* represent the margins of profitability. Profitable production can be anywhere between *Ma* and *Mb*. Figure 19b represents the converse of Fig. 19a. Here costs are assumed to be constant, but demand, and therefore price also, is assumed to vary from area to area. The optimum location is still at 0 and the margins at *Ma* and *Mb*. Hence, the concepts of optimum and marginal location apply in the same way whether demand or supply is held constant. In the real world both cost and demand have areal variations. In Fig. 19c, costs rise with distance from point *A*, while demand, reflected in price, is greatest at point *B*. At *A*, costs are lowest, and this is the maximum profit point, the optimum location, despite the fact that demand is greatest at *B*. The entrepreneur seeking maximum profits would locate at *A* despite the lower total revenue obtainable there. By altering the gradients of the cost and price lines the optimum location could become the point of highest revenue.

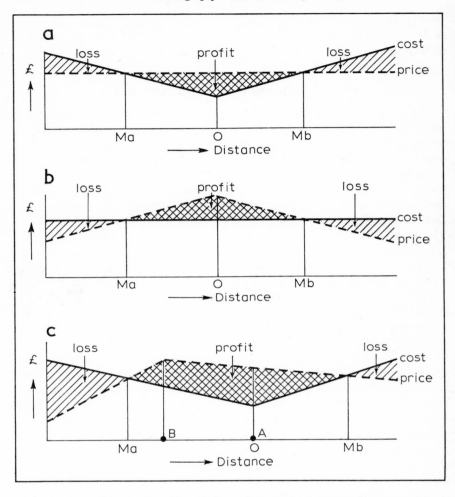

Fig. 19. Theoretical models of industrial location. (After Smith, 1966.) For explanation, see text.

TESTING LOCATION THEORY

Much effort has been expended by economic geographers and economists in developing and refining a body of theory for use in location analysis. Even more effort has been put into empirical studies of various types of industrial activity and the areas in which they are located. These have been mainly to gain a general understanding of the factors influencing location and the resulting patterns, rather than to test theory. There are two reasons for this.

First, there is the difficulty of generating testable hypotheses from an existing theory which contains so many simplifying assumptions that it is almost completely divorced from the real world. In Smith's (1966) model, for example, it is assumed that the sources of such production factors as labour and materials are fixed, that their supply is unlimited, and that substitution between them is impossible. Weber himself assumed that both demand and price are constant in space and that the location decision will rest solely on supply costs. In addition, not only are methods of analysing location patterns and their variations still poorly developed—although Isard's (1960) contribution in this field has been particularly notable—but the necessary data are only now becoming available, owing to improved techniques of collation and processing. A second, and perhaps more significant reason why the testing of theory has been given little attention is that the goals of theoreticians and empirical researchers have been different. The former have been more concerned with defining optimum spatial patterns and explaining divergencies from them, while the latter have been more pragmatic, being interested in the implications of manufacturing patterns for problems of economic development, city growth, and regional planning.

Of the few attempts to test theory, one of the most significant is that of Smith (1955). Taking Weber's proposition that those industries with a great loss of weight during manufacture will tend to be located close to their raw materials, he tried to determine its relevance to the observed facts of location in Britain. He developed a loss of weight ratio (weight of material/weight of product) which enabled him to compare data for sixty-five industries in the Census of Production. He found that only those industries with high weight-loss ratios had a close relationship with raw materials. There were many industries in the middle ranges of weight-loss whose location was only partly controlled by their raw materials. Refining the theory further, Smith developed an index based on weight of materials per operative; if weight of material handled per operative was high the industry would tend to be located close to its raw materials. Introduction of this new factor, similar to Weber's use of labour costs, provided a more effective tool and the area of uncertainty was narrowed. Smith's test enabled him to echo Weber in concluding that "loss of weight has significant locational effects only when it is combined with large weight per operative, for variations in transport costs are substantial enough only if weights handled are large" (Smith, 1955, p. 11). Smith's paper was merely a beginning for it was concerned only with evidence relating to weight of materials used in the manufacturing process. His untimely death prevented him considering the vast concourse of industry not located by materials. Regrettably, his work has not been developed by other geographers.

Another pioneering attempt at testing theory was that of Harris (1954). He aimed to incorporate both population potential, as a measure of market size, and transport cost minimization into the analysis of location patterns. The market potential (P) he assessed as

$$P = \sum \frac{M}{d},$$

i.e. the sum of markets accessible to a point (M) divided by their distances (d) from that point. Distance was represented by transport costs. Harris considered the market in national terms and demonstrated that in the United States the areal distribution of market potential shows the highest concentrations in the manufacturing areas of the northeast. He took iron and steel as an example, illustrating the decreasing significance of the location of raw material sources in manufacturing. Pittsburgh, for example, was shown to have a "superb central location amidst the huge markets of the American Manufacturing Belt" (Harris, 1954, p. 318).

Studies such as those of Smith and Harris are few and the methods of analysing location patterns are still poorly developed. Much methodological research has been directed towards the problem of measurement, especially of locational shifts of industrial growth and its secondary impacts. The localization of industry index, first developed by Sargent Florence (1948) in a notable study of the relationships among size of plant, investment, and location, has been a favourite measure. He developed a "location quotient"—a device for comparing a region's percentage share of a particular activity with its percentage share of some basic aggregate such as employment or income—as a measure of regional industrial concentration. Using this he was able to correlate size of plant with degree of localization, and capital intensity of the production process with plant size. Location quotients have since been used in studies of regional economies and have provided convenient summaries of the general location characteristics of particular areas. The main disadvantage of this sort of measure is its use of the national average as a norm, implying something inherently correct about one area having a similar industrial structure to the nation as a whole. The national picture is, of course, merely the sum of several regional patterns, each of which may result from comparative advantages.

EMPIRICAL STUDIES

Most literature on industrial location consists of empirical case studies of locational factors such as labour and materials, of selected industries, and of regions. Their aim is to provide some insight into how locational factors work and how their influence varies from industry to industry and area to area. For example, there have been many studies of the transport factor, and the importance of accessibility to communications has been frequently explored (Alexander *et al.*, 1958; Fulton and Hoch, 1959). Labour as a factor of production has been a common theme, although concentration has centred on the issue of wages (Segal, 1960), while the equally important factors of physical availability of manpower, and especially of manpower quality, have been largely ignored. A growing body of literature deals with problems of both internal and external economies of scale (Lomas and Wood, 1969). The effects of changing technology will obviously need much attention, while automation, which increases the capital input in manufacturing while reducing the labour input, is bound to have an important influence on location decisions.

Other empirical studies have put emphasis on specific industries and the factors affecting them. For example, Isard's (1948) study of the iron and steel industry has shown how changing material inputs in the production process in the form of increased scrap and less pig iron used, can increase the pull of the market (cf. Harris, 1954). More recently, Manners (forthcoming) has produced a comprehensive analysis of the location of iron and steel industries on an international scale. Other workers have used selected industries to demonstrate aspects of location theory. Hoover (1937) used the shoe and leather industries in this way, while Estall (1963) used the electronics industry to note linkages between production and research and development facilities.

Empirical location studies have taken place in the context of specific areas, with attempts to explain the character of existing industrial structures. Martin's (1966) study of industrial London and Smith's (1953) study of the location of industry on Merseyside come into this category, and a volume on South Wales (Manners, 1964) represented the attempt of a group of specialists to analyse the problems of one industrial region. The development of specialized industrial "quarters" within cities has been analysed by Wise (1949) and Hall (1962). Regional "shifts" of manufacturing have been studied by Zelinksy (1958) in the United States, while Keeble (1968) has investigated industrial movement in northwest London.

THE SIGNIFICANCE OF ENTERPRISE

Thus far, study of industrial location has been primarily concerned with its logic. It must, however, be recognized that the individual entrepreneur may have his own ideas of what constitutes an ideal location. Traditional location theory has been mainly concerned with minimizing transport costs, and optimum location has usually been defined in these terms. But there are other considerations in the location decision, many of which are difficult or even impossible to quantify. A manufacturer may prefer to locate close to his favourite golf course; provided he can make a profit, he will be satisfied. More important are problems of uncertainty and time. Technology and tastes may change so that what is practical in an economic sense today may mean low profits or loss 5 years hence. The result is that often there is no single best strategy for the entrepreneur. He may gamble, in the hope of big profits, or act guardedly, preferring small but safe profits. Both are rational decisions and the choice depends on his aims and attitudes.

In recent years a few geographers have become concerned with the human element in location decisions, and at the same time have criticized the idea of optimum location (Rawstron, 1958). Pred (1967) has collected together ideas, such as those of March and Simon (1958), on the effects of behaviour on location. These claim that most entrepreneurs are content with locations that seem to provide them with acceptable levels of profit rather than seeking out the optimum. Interesting as these ideas are, they do not detract from the conceptual value of the optimum location. Their main significance is that they focus attention on the importance of enterprise in considering industrial location.

The increasing emphasis in economics on the theory of the firm means that industrial geographers will undoubtedly have to study the geography of the firm, considering not only man himself but also his social organizations and institutions. One of the basic weaknesses of location theory has been its concern almost solely with production units, whereas modern manufacturing firms frequently consist of spatially separate but locationally interdependent administrative, research, development, sales, and production units. McNee (1960) argued that the most important single institution in the urban–industrial complex is the modern corporation. Policy decisions taken at Head Office can affect thousands of workers at many locations. Within a corporation framework, a man may migrate from plant to plant. For example, the new plant of Vauxhall Motors Ltd. at Ellesmere Port resulted in the transfer there of several hundred employees from the company's existing factories in Luton and Dunstable. Such a migration from one point in a company's regional system to another point in the same system is planned and often financed by the company itself.

The significance of the rise of the modern corporation, with its attendant "technostructure", has been chronicled by Galbraith (1967, p. 1):

> Seventy years ago the corporation was still confined to those industries—railroading, steamboating, steel-making, petroleum recovery and refining, some mining—where, it seemed, production had to be on a large scale. Now it also sells groceries, mills grain, publishes newspapers and provides public entertainment, all activities that were once the province of the individual proprietor or the insignificant firm. The largest firms deploy billions of dollars' worth of equipment and hundreds of thousands of men in scores of locations to produce hundreds of products. The five hundred largest corporations produce close to half of all the goods and services that are available annually in the United States.

Such changes in economic organizational structure account for the growing interest in the geography of the firm. The atomistic economy of the nineteenth and early twentieth centuries is rapidly giving way to the economy of giant industrial combines. It is easy to comprehend that the regional policies of these combines can be decisive in the regional patterns of many small accessory manufacturers, although lack of data makes their study difficult.

The industrial geographer must, if he is to succeed in explaining location patterns, be aware of the significance of structural changes in the economy and be able to interpret them. He must study the characteristics of the new economy and its corporations. First and foremost he must be aware that spatial decision-making is no longer only the result of the testing of many individual decisions, but of highly complex pre-formulated plans. Furthermore, such plans are prepared and implemented by a specialist managerial class whose business is to take decisions. In the modern corporation, ownership (shareholders) and control (managers) are now separated (McNee, 1960). Second, the growth of corporate regional planning has to be recognized. The value of a particular location depends not so much on its own inherent advantages and disadvantages of site and position, but on its corporate position. The decision to open or close any one production unit may depend less on its own cost structure than on how it fits into the overall corporate strategy. A third consideration is the continuing rôle of government, either as owner of public corporations itself or as an arbitrator between the great private corporations through agencies like the Industrial Reorganization Corporation in the United Kingdom.

These developments represent considerable progress from the geography of price, the traditional concern of industrial location. There are good grounds for agreeing with McNee (1960, p. 206) that "in the large scale corporate regional systems of our time the geography of price has become fused with the geography of administration". Study of specific corporate regional systems would seem to be one of the most fruitful avenues of investigation for the industrial geographer.

REFERENCES AND FURTHER READING

A review and annotated bibliography of theoretical, empirical and case studies in industrial ocation is:

STEVENS, B. H. and BRACKETT, C. A. (1967) Industrial location, *Regional Science Research Institute Bibliography Series*, **3.**

Among general works on the subject is:

ALEXANDERSSON, G. (1967) *Geography of Manufacturing*, Prentice-Hall, New Jersey.

The need for a theoretical structure is discussed in:

CHISHOLM, M. (1966) *Geography and Economics*, Bell, London.
SMITH, W. (1955) The location of industry, *Transaction of the Institute of British Geographers*, **21,** 1–18.

LOCATION THEORY AND THE CONCEPT OF OPTIMUM LOCATION

Location theory is developed and reviewed in numerous works. Early ones are:

MILL, J. S. (1848) *Principles of Political Economy.*
RICARDO, D. (1817) *The Principles of Political Economy and Taxation.*
SMITH, A. (1776) *Inquiry into the Nature and Causes of the Wealth of Nations.*

Modern location theory is based on:

GREENHUT, M. L. (1956) *Plant Location in Theory and Practise: the economics of space,* U. of N. Carolina P., Chapel Hill.
HOOVER, E. M. (1948) *The Location of Economic Activity,* McGraw-Hill, New York.
ISARD, W. (1956) *Location and Space Economy,* M.I.T.P., Cambridge, Mass.
LÖSCH, A. (1940) *The Economics of Location,* English translation 1954, Yale U.P., New Haven.
WEBER, A. (1909) *Theory of the Location of Industries,* English translation 1929, U. of Chicago P., Chicago.

THE SIGNIFICANCE OF COST AND PRICE

A simple model for industrial location is built in:

SMITH, D. M. (1966) A theoretical framework for geographical studies of industrial location, *Economic Geography,* **42,** 95–113.

The concept of optimum location is attacked in:

RAWSTRON, E. M. (1958) Three principles of industrial location, *Transactions of the Institute of British Geographers,* **25,** 135–42.

TESTING LOCATION THEORY

The best works on methodology are:

ISARD, W. (1960) Methods of regional analysis, *Regional Science Studies, series 4,* M.I.T.P., Cambridge, Mass.
SARGENT, F. P. (1948) *Investment, Location and Size of Plant,* Cambridge U.P., Cambridge.

In addition to the case study of Smith (1955), see also:

HARRIS, C. D. (1954) The market as a factor in the location of industry in the U.S., *Annals of the Association of American Geographers,* **44,** 315–48.

EMPIRICAL STUDIES

Among works which discuss the importance of aggregation and linkage is:

LOMAS, G. and WOOD, P. A. (1969) *Employment Location in Regional Economic Planning,* Cass, London.

Wages as a location factor are discussed by:

SEGAL, M. (1960) *Wages in the Metropolis: their influence on the location of industries in the New York region,* Harvard U.P., Cambridge, Mass.

The transport factor is discussed in:

ALEXANDER, J. W., BROWN, S. E. and DAHLBERG, R. E. (1958) Freight rates: selected aspects of uniform and nodal regions, *Economic Geography,* **34,** 1–18.

FULTON, M. and HOCH, L. C. (1959) Transportation factors affecting location decisions, *Economic Geography*, **35**, 51–9.

Studies of specific industries include:

ESTALL, R. C. (1963) The electronic products industry of New England, *Economic Geography*, **39**, 189–216.
HOOVER, E. M. (1937) *Location Theory and the Shoe and Leather Industries*, Harvard U.P., Cambridge, Mass.
ISARD, W. (1948) Some locational factors in the iron and steel industry since the early nineteenth century, *Journal of Political Economy*, **56**, 203–17.
MANNERS, G. (forthcoming) *The Changing World Market for Iron Ore, 1950–1980*.

Case studies of industrial regions include:

HALL, P. (1962) *The Industries of London Since 1861*, Hutchinson, London.
MANNERS, G. (Ed.) (1964) *South Wales in the Sixties*, Pergamon, Oxford.
MARTIN, J. (1966) *Greater London: an industrial geography*, Bell, London.
SMITH, W. (1953) The location of industry on Merseyside, *Tijdschrift voor Economische en Sociale Geografie*, **44**, 101–9.
WISE, M. J. (1949) On the evolution of the jewellery and gun quarters in Birmingham, *Transactions of the Institute of British Geographers*, **15**, 57–72.

Among analyses of industrial movement are:

KEEBLE, D. E. (1968) Industrial decentralization and the metropolis: the North-west London case, *Transactions of the Institute of British Geographers*, **44**, 1–54.
ZELINSKY, W. (1958) A method of measuring change in the distribution of manufacturing activity: the United States, 1939–47, *Economic Geography*, **34**, 95–126.

THE SIGNIFICANCE OF ENTERPRISE

One of the best short works in this field is:

McNEE, R. B. (1960) Towards a more humanistic economic geography: the geography of enterprise, *Tijdschrift voor Economische Sociale Geografie*, **51**, 201–6.

More comprehensive is:

GALBRAITH, J. K. (1967) *The New Industrial State*, Hamish Hamilton, London.

Works on the importance of behaviour in location decision include:

MARCH, J. G. and SIMON, H. A. (1958) *Organizations*, Wiley, New York.
PRED, A. (1967) Behaviour and location, *Lund Studies in Geography, series B*, **27**.

CHAPTER 18

THE DIVERSITY OF URBAN GEOGRAPHY

J. H. JOHNSON

URBAN geography is not only a very rapidly growing field of study, but it is also a remarkably diverse one, with the interests of urban geographers varying from time to time almost with the caprice of fashion. As well as being of increasing interest to research workers, courses on urban geography are popular among undergraduates and the subject is now beginning to penetrate the work of secondary schools. Yet in spite of its blossoming success, it is fair to say that urban geography does not possess a definitive methodology: its unity is provided by the subject matter studied, rather than by any single method of approach. As a result it may be helpful to isolate some of the themes which urban geographers have studied at various times and to comment on the advantages and disadvantages which these different approaches offer.

THE STUDY OF URBAN MORPHOLOGY

There has been a long-standing preoccupation among geographers with the morphology of cities, or the "townscape", as some have styled it. Many geographers, particularly those working in Germany, have taken the classification and mapping of distinctive morphological areas within cities as an end in itself, and, indeed, some have emphasized the actual appearance of cities. In Britain, for example, Smailes (1955) has noted the "block clumping", "ribbing", and "studding" which form the urban texture of many cities, and has mapped these features in east London (Smailes and Simpson, 1958). Similarly, Johns (1965) has also been primarily interested in the appearance of the urban scene, his work being concerned with the contribution of different architectural periods and styles to the shaping of British cities in general. A problem is presented by the great variety of scales at which work of this kind can be undertaken, each with their own limitations. Studies of urban morphology can range from an examination of small groups of buildings, concerned with how they fit into the appearance and functioning of a local area, to the division of whole cities into broad morphological regions.

Yet in spite of the appeal of urban morphology for geographers, there have been few studies which make a comprehensive analysis of the urban fabric of large individual cities. The reasons for this absence are not difficult to find. The morphology of a city is made up of a large number of components, which include the urban plan, the nature of the buildings which line the streets, and, not least, the physical expression of the functions which dominate different parts of the city (Conzen, 1960 and 1968). This variety of components impedes easy description by map or by word, and explanation is complicated by the wide range of forces, often independent of one another, which influence urban morphology. Any comprehensive study of the bricks and mortar of a city must consider, among other things, the history of architecture, the process of urban growth, the various functions which different parts of the city undertake, and the social attitudes of the various classes of people who inhabit it.

Because of this complexity the most successful studies of urban morphology have been those that limit the aspects of the urban fabric with which they are concerned and, at the same time, attempt to link morphology with a restricted range of other phenomena. For example, Prince's study of the growth of north-west London is a personal view of the area, concerned with the interplay among landowners, railway building, and the nature of the social groups which came to occupy new residential areas (Prince, 1964). Or again, Vance (1966 and 1967) has examined the urban structure of Birmingham in terms of the location of industrial plants and the related housing of factory workers.

Critics of the study of urban morphology have stressed its descriptive nature, which it is claimed does scant justice to the complex linkages among the various geographical phenomena within cities. Certainly morphology is merely one expression of urban geography, and the approach has the danger that it can degenerate into a collection of unstructured information, which may have local topographical interest, but which lies outside the mainstream of intellectual interest in urban matters. Yet the selective quality which the more successful of these studies possess implies that the descriptions which they offer are rarely "pure", since even description requires certain assumptions about what is valid and relevant (Klimm, 1959).

The important point, stressed all too clearly by the failings of many undergraduate dissertations and school projects as well as some research work, is the need for an underlying problem or, at least, a clearly-defined purpose. Given this, urban morphology remains a valid academic concern at a variety of levels. Rasmussen's description of London, although not always a reliable guide to the detailed facts of architectural history, remains a classic because of its idiosyncratic but perceptive account of the architectural fabric of the metropolis (Rasmussen, 1937). At a more detailed scale and with quite a different flavour, P. A. Stone's analysis of the building costs and residential densities associated with tall flats and conventional houses has given important meaning to one aspect of urban morphology (Stone, 1967). Clearly, the need for selection in studying urban morphology is not simply an unwelcome necessity, but is essential for the intellectual vigour of work in this field.

While the morphology of an individual city may have many unique features, it will also bear a family resemblance to other cities which have grown up at the same time and in similar economic and cultural contexts. Hence a final problem in urban morphology is provided by the need to establish generalizations relevant to different cities. Although one of the attractions of detailed studies of urban morphology is the appreciation which they yield of the distinctive texture of particular urban areas, it is necessary to fit these observations into general models of urban types.

Two lines of approach have commonly been adopted. One is to emphasize the process of historical development in shaping similar townscapes, an approach which is found more often in the work of European geographers. In Britain, for example, Fleure (1936) and Dickinson (1951) have stressed the cultural and historical similarities which emerge in the morphology of West European cities. A second approach emphasizes the different functional regions found within cities, which also find expression in urban morphology. Studies by American geographers, for instance, have usually shown this interest, presumably because of the relatively recent growth and relatively homogeneous cultural history of North American cities. Their fundamental interest, however, has been in the areas used for different functions, rather than in the nature of buildings and street patterns, so that it is reasonable to see work of this kind as a separate strand in the interests of urban geographers.

URBAN LAND-USE REGIONS

The expansion of cities over the last 100 years and the development of more efficient methods of internal transportation within urban areas has led to the development within cities of extensive areas which possess specialized land uses and functions. These regions include the city centre, industrial areas, outlying commercial areas, residential areas, and the urban fringe where rural and urban uses are intermixed. The study of these regions has become an important theme in urban geography.

One element in this research is a concern with the actual areas of land used for particular urban purposes, since information of this kind is of immediate value for planning purposes and for the formation of soundly based opinions about policies to be adopted for urban areas. Pioneer work in America was undertaken by Bartholomew (1932 and 1955), who made a comparative study of cities to assess the amount of land required for various purposes in typical American cities, hence providing a guide to the allocation of land for future urban expansion. More recently, much attention has been given in Britain to mapping the land-use patterns found in individual cities, since surveys of this kind are required in the preparation of official development plans (Keeble, 1964).

Modern studies of urban land uses have become more focused on specific problems, particularly those found around the fringes of urban development. Thomas (1964), for example, has investigated land use in London's Green Belt, where the general economic and social processes influencing the use of land in an urban fringe area have been operating in the particular context of a strictly administered planning framework. Similarly, Best's assessment of the extent of recent urban growth in Britain has explored a most important matter for such a highly urbanized country; and, incidentally, this work has revealed remarkable official ignorance of the rate of urban growth, even though it is a topic of long-standing controversy (Best, 1968).

Land uses have also been used to delimit urban regions in order to allow their further study. In particular, work on the delimitation of central business districts has frequently used land use information as a most important criterion (Murphy and Vance, 1954; Davies, 1960), although land-use studies in city centres are complicated by the three-dimensional quality of these areas, with many buildings housing a variety of uses at different floor levels. The current use of land has also been used to indicate past stages in urban growth (Whitehand, 1967), to establish the detailed structure of shopping centres, and to identify the various concentrations of retail and service establishments which form the general pattern of retail business within cities (Cohen and Lewis, 1967).

In recent years, however, geographers have become more concerned with the processes that lie behind the formation and characteristics or urban land-use regions than with the detailed delimitation of their boundaries. In each urban region there is a series of relationships between various economic and social factors which shape its nature; and all the various regions within a city are similarly linked together by the movement of people and goods. One of the most obvious of these links is the journey to work, connecting place of residence with place of employment, and hence being a strong underlying force behind the development of the most important urban regions. Studies of this daily movement are numerous; and geographers are now exploring not only movements within the built-up areas of cities, but also the manner in which urban employment is extending its influence into apparently rural areas (Lawton, 1963). Today the motor car allows urban workers to live in the countryside as conveniently as in the town, thus blurring at least some of the distinctions between urban and rural society.

Studies of the linkages found between separate enterprises in industrial and commercial areas and of the use of recreational land by urban dwellers also stress the processes which lie behind land use (Burton and Wibberley, 1965; Goddard, 1968; Starkie, 1967). Similarly, studies or urban transport have been stimulated by the same approach. Research funds, particularly in the United States, have been most readily available for the study of road transport; and numerous investigations have examined the impact of road developments on other features of the city (Garrison *et al.*, 1959). Work has also been undertaken on this problem from the other end and has resulted in a monumental, and as yet unintegrated literature on the manner in which various land uses themselves generate traffic (Boileau, 1959; Medhurst, 1963).

An awareness of the linkages present in the city has also influenced studies of the growth of urban areas. This work is particularly important in that it allows the development of statistical models of urban growth, which seek to define the interconnections between the various factors involved in this process and, if possible, to quantify them. Such models, backed by the capacity of electronic computers to handle vast quantities of data, allow informed predictions of future urban growth (Cowan *et al.*, 1967; Hall, 1967; Morrill, 1965). These simulations are probably most successful as an indication of relatively short-term developments, although it is difficult to assess the precise length of time over which they are likely to produce their best results. Predictions of this kind would be destroyed by any radical innovation in urban transportation, and over a long period it is difficult to assess the changing attitudes of society to the form and rate or urban growth. These simulations, however, represent an important statement of what the future may hold if present trends are continued.

THE HUMAN ECOLOGY OF CITIES

Closely connected with the work of geographers on urban land-use regions is their interest in what has been called the human ecology of cities. This approach is stimulated by the belief that the most important feature of the city is the society which has given rise to it, occupies it, and shapes its form; and it is an approach which sprang originally from the work of sociologists. Studies of the human ecology of urban areas date from the flowering of the Department of Sociology of the University of Chicago in the 1920's and 1930's. R. E. Park, the driving force behind this department, held " . . . a conception of the city, the community and the region, not as a geographical phenomenon really, but as a kind of social organism" (Park, 1952, p. 5). As a result, much of the work of Chicago sociologists was concerned with the investigation of human behaviour in an urban environment.

An approach of this kind has obvious attractions for geographers, since an interest in the interactions between human societies and their environments has always been a repetitive theme in much geographical writing. Over the years the research of sociologists and geographers has tended to follow rather different paths. Much of the classic work by the Chicago school of urban sociologists examined small groups of people, and was particularly interested in such topics as their assimilation and social integration. Geographers, on the other hand, have emphasized the spatial aspects of the ecological processes found within whole cities. In practice much of their work in this field has examined some characteristics of the population which occupies the various morphological and functional regions of particular cities (e.g. Thomas, 1960; Carey, 1966).

In theory, at least, the attraction of urban areas for studies of human ecology is enhanced because here the natural environment does not attain the same level of importance as in a rural setting. As a result, generalizations about the spatial expression of human behaviour are likely to emerge more clearly and repetitively in urban areas. Possibly this is one of the reasons why this approach to urban geography should have regained more attention in recent years, as human geography as a whole has attempted to move away from the particularism provided by individual areas towards wider generalizations. One indication of this revived interest is the reprinting of Burgess's famous diagram of the concentric structure of the social regions within Chicago in a number of modern textbooks, where it has been elevated to the status of a model (Haggett, 1965, pp. 177–8). Curiously enough, Burgess's statement is now over 40 years old and, whatever the validity of the concentric theory of urban structure, the urban form which he was describing was that of the American city of the 1920's (Park *et al.*, 1925, pp. 51–3).

Two practical factors have also encouraged renewed interest in the human ecology of cities. One element is the availability of census data by enumeration districts; the other is the application of computers to the handling of data. As a result it has now become possible to analyse a considerable amount of detailed information concerning the population of cities, and also to quantify the various relationships which exist between such diverse features as social class, housing quality, and demographic structure. What have become standardized procedures like component and cluster analysis can be applied to grouping this information to form statistically meaningful regions (Robson, 1966).

In a sense the application of more rigorous statistical techniques to an immensely large body of data may represent an advance, since it means that general statements can be built up from detailed information covering a whole city of cities (Gittus, 1964). On the face of it such an approach is more attractive to geographers than the original pioneer studies by sociologists in Chicago, which were often concerned with specialized groups of people like hobos and gangs. Yet, at the present stage of geographical work, it is doubtful if an important intellectual stride has in fact been taken. The data most commonly used is the socially incomplete collection made in census returns, and there is no assurance that the most important variables have been included in these analyses. Although it is now possible to map the patterns of human ecology within individual cities in much greater detail than was possible before, it is by no means certain that these inductive methods will produce better generalization about the social patterns within cities; nor is there any clear sign that they are revealing much about the processes which produce the patterns revealed. The original work in Chicago applied patient fieldwork methods, already developed by social anthropologists, to problems of urban life and culture; but it can be argued that recent geographical work has not produced comparable probing insights into the functioning of urban society, which must be the basic aim of work of this kind. Descriptions have become more detailed, but explanations have not been equally deepened.

A reaction to the great mass of data which has now been assembled for a diverse range of cities is being provided by what has been called "gradient analysis". This approach springs from the observation that various important features of the geography of cities change in a logical sequence on a traverse from city centre to urban fringe. As a result graphs can be constructed to illustrate this progression and equations calculated to describe the gradients revealed on the graphs. Such work is not new, but in recent years there has been revived interest in the application of this technique to the analysis of urban population densities (Clark, 1951). In part this interest has been stimulated by the parallel work of land economists, who have recently been giving more attention to the operation of the residential sector of the urban land market, which provides a partial explanation of the density gradients which are commonly found (Alonso, 1965).

The great advantage of this approach is the manner in which it simplifies the complex patterns within cities by reducing them to relatively simple mathematical statements, thus making it possible to compare different cities. These comparisons have established that population densities decrease at a constant rate with increasing distance from city centres; and this relationship has been observed in a wide range of cities, with a variety of locations. The same technique allows comparisons of the changing population gradients found at various times in the past in individual cities. In particular a contrast has been made between western cities, where gradients have decreased since the last quarter of the nineteenth century, and non-western cities where densities as a whole are thought to have increased but gradients are believed to have remained stable (Berry *et al.*, 1963). Assumptions about the trend of population gradients have also been used to project the future pattern of population distribution within cities (Newling, 1964).

CITIES AS CENTRAL PLACES

Another important theme which has been occupying the attention of urban geographers is the analysis of the role of towns and cities as central places, that is settlements concerned with the provision of goods and services, a function which includes administrative, financial, and other similar services, as well as retail and wholesale trade. These businesses require a location with maximum accessibility to the people who use them and hence tend to cluster in market centres visited by surrounding consumers. But services have various levels of specialization and those central places which can tap a larger number of potential customers than their rivals tend to possess more specialized services and grow larger than the others. The resulting system of central places of various sizes exhibits a degree of regularity, which has encouraged geographers and others to formulate theoretical generalizations about the distribution and size of central places.

Again, this work is not new. In 1933, Christaller formulated a number of theoretical arrangements of central places, the most important of which demonstrated what he called the "marketing principle", with the number of centres at successively less specialized levels following a regular geometric progression with a factor of 3 (Christaller, 1933 and 1966). Christaller's work was introduced to English-speaking geographers by Ullman (1941) and Dickinson (1947). More complex theoretical arrangements were suggested in 1940 by Lösch and became accessible to British and American geographers in the 1950's after the translation of Lösch's book into English and the parallel attempt by Isard to integrate much of the literature in this field (Lösch, 1954; Isard, 1956). As a result, considerable effort since the late 1950's has been devoted to the assessment of these theoretical schemes and to the testing of their validity in real situations.

Most attention has been given to the work of Christaller, particularly after his ideas had been refined by Berry and Garrison (1958). Berry and Garrison emphasized the importance of the "threshold" (the minimum amount of purchasing power necessary for a particular service to survive) and the "range of a good" (the distance which people are prepared to travel in order to obtain a particular service); and they described how the interaction of these two factors produced a hierarchy of service centres, thus giving a theoretical justification for a phenomenon which was already well known from empirical studies (e.g. Smailes, 1946; Brush, 1953).

Central-place theory has made a number of other important contributions to urban geography. For one thing, it has provided a partial explanation of the presence of a varying number of settlements of different sizes, with a large number of small towns and a smaller number of larger settlements. In many areas this arrangement roughly follows what has been called the "rank-size" rule, which states that, if the urban settlements in an area are ranked in descending order of population size (from 1 to n), the population of the nth town will be $1/n$ the size of the largest city. There is a broad similarity between the rank-size arrangement of city sizes and that produced by Christaller's central place hierarchy, provided that the presence of a random element is accepted to blur the edges of the hierarchial system of central places (Beckman, 1958). As a result it has been argued that central-place theory provides a most convincing explanation of why the population of cities follow the rank-size rule (Bunge, 1962). If this view is correct (and the matter is still open to debate), it stresses the importance of the function of cities as central places in controlling their general size.

Central-place theory also makes an important contribution to the study of the numerous shopping centres within those cities which are large enough to have a complex pattern of retail distribution. Early studies in Philadelphia by Proudfoot (1937) revealed a hierarchy of shopping centres, including small clusters of stores, neighbourhood business streets, principal shopping thoroughfares, and outlying business centres, as well as the central business district. Although this particular retail structure must now be somewhat recast because of the growth of new shopping centres focused on the motor car and sometimes serving whole sectors of large cities (Vance, 1962), the hierarchy of shopping centres provides a parallel, if not an exact one, to the hierarchy of towns and cities of different sizes and levels of retail specialization. The "range of a good" and the "threshold" still stand as important concepts in the analysis of the functioning of these centres, although the rise of metropolitan centres, with more dispersed urban structures and with the increased flexibility of movement brought by the motor car, may lead to a decline in the validity of these concepts (Berry, 1967, p. 124). This, however, is a matter for the city of the future, rather than part of the present urban scene.

A fully developed central-place system with a hierarchy of regularly spaced settlements of various sizes is rarely, if ever, found in reality. The value of these theories, however, is enhanced by the fact that consideration of a theoretical situation, which assumes an even distribution of population and perfect competition between urban centres, allows a clearer understanding of the social and economic relationships which exist between town and country and between towns at various levels in the hierarchy, without the complications provided by the uniqueness of every place on the earth's surface. In addition, the discrepancies which can be found between model and reality pose problems for further investigation.

THE NEED FOR A CULTURAL CONTEXT

It is a fair criticism of urban geography that the various approaches to the subject which have been described here have largely been applied to the study of western cities, and that most of them use techniques which have been devised with western conditions in mind. Because of the sheer importance of urban life in western society, it is likely that this situation will continue in the future, particularly as non-western cities become increasingly influenced by western culture.

In recent years, however, a number of urban geographers have turned their attention to non-western cities. Although sometimes their work has suffered from the application of techniques which are valid in western society, but not necessarily so elsewhere, there are signs that the particular social context is being increasingly assessed in those urban studies which venture into different culture worlds. For example, a literature is being built up which seeks to describe the mixture of urban types found in various regions (e.g. McGee, 1967; Horvarth, 1968), and which examines individual cities with due regard to their social background (Mabogunje, 1962). Attempts are also being made to adapt well-established theories to conditions which exist outside the west. For instance, central-place theory is being interpreted in the context of traditional societies, to allow for such features as the rôle of travelling peddlers in such communities (Skinner, 1964; Berry, 1967, pp. 89–105).

In short, data is now being collected which will allow some of the general assumptions of urban geography to be tested in different cultural contexts. The next stage of this work will demand the cross-cultural comparison of cities; but even where attention is focused on urban settlements lightly touched by modern commercial and industrial growth, there are grave problems in this task (Sjoberg, 1960; Wheatley, 1965). What remains clear is that, while geographers may be able to take the *mores* of their own society for granted, the social and economic assumptions which are valid in western society do not necessarily apply elsewhere—an observation which applies to the whole content of human geography and not just to the study of towns and cities.

REFERENCES AND FURTHER READING

URBAN MORPHOLOGY

British studies of urban morphology include:

CONZEN, M. R. G. (1960) Alnwick, Northumberland: a study in town plan analysis, *Transactions of the Institute of British Geographers*, **27**.

CONZEN, M. R. G. (1968) The use of town plans in the study of urban history, in DYOS, J. H. (Ed.) *The Study of Urban History*, Arnold, London, pp. 113–30.

JOHNS, E. (1965) *British Townscapes*, Arnold, London.
SMAILES, A. E. (1955) Some reflections on the geographical description and analysis of town-scapes, *Transactions of the Institute of British Geographers*, **21,** 99–115.
SMAILES, A. E. and SIMPSON, S. (1958) The changing face of east London, *East London Papers*, **1,** 31–46.

More specialized studies are:

PRINCE, H. C. (1964) North-west London, 1814–1863; North-west London 1864–1914, in COPPOCK, J. T. and PRINCE, H. C. (Eds.) *Greater London*, Faber, London, pp. 80–141.
RASMUSSEN, S. (1937) *London, the Unique City*, Cape, London. (This book was reprinted in an abridged form by Penguin Books in 1960.)
STONE, P. A. (1967) Urban form and resources, *Regional Studies*, **1,** 93–100.
VANCE, J. E., JR. (1966) Housing the worker: the employment linkage as a force in urban structure, *Economic Geography*, **42,** 294–325.
VANCE, J. E., JR. (1967) Housing the worker: determinative and contingent ties in nineteenth century Birmingham, *Economic Geography*, **43,** 95–127.

The problem of making "pure" descriptions is explored briefly in:

KLIMM, L. E. (1959) Mere description, *Economic Geography*, **35,** facing 1.

Features of the morphology of European cities described in:

DICKINSON, R. E. (1951) *West European City*, Routledge, London.
FLEURE, H. J. (1936) The historic city in western and central Europe, *Bulletin of John Rylands Library*, **20,** 312–31.

URBAN LAND-USE REGIONS

Land-use studies which exhibit a variety of approaches include:

BARTHOLOMEW, H. (1955) *Land Uses in American Cities*, Harvard U.P., Cambridge, Mass. Originally published in 1932 as *Urban Land Uses*, Harvard U.P., Cambridge Mass.
BEST, R. (1968) Extent of urban growth and agricultural displacement in post-war Britain, *Urban Studies*, **5,** 1–23.
KEEBLE, L. (1964) *Principles and Practice of Town and Country Planning*, 3rd edn., Estates Gazette, London, chapter 9.
THOMAS, D. (1964) The components of London's green belt, *Journal of the Town Planning Institute*, **50,** 434–9.

For the use of land-use information to delimit other urban features see:

COHEN, S. B. and LEWIS, G. K. (1967) Form and function in the geography of retailing, *Economic Geography*, **43,** 1–42.
DAVIES, D. H. (1960) The hard core of Cape Town's central business district: an attempt at delimitation, *Economic Geography*, **36,** 347–65.
MURPHY, R. E. and VANCE, J. E., JR. (1954) Delimiting the C. B. D., *Economic Geography*, **30,** 189–222.
WHITEHAND, J. W. R. (1967) Fringe belts: a neglected aspect of urban geography, *Transactions of the Institute of British Geographers*, **41,** 223–33.

For examples of studies which stress the movement of people and goods and other linkages influencing land use see:

BOILEAU, I. (1958) Traffic and land use, *Town Planning Review*, **29,** 27–42.
BURTON, T. J. and WIBBERLEY, G. P. (1965) *Outdoor Recreation in the British Countryside*, Wye College, Ashford.
GARRISON, W. L., BERRY, B. J. L., MARBLE, D. F., NYSTUEN, J. D. and MORRILL, R. L. (1959) *Studies of Highway Development and Geographic Change*, U. of Washington P., Seattle.
GODDARD, J. (1968) Multivariate analysis of office location pattern in the city centre: a London example, *Regional Studies*, **2,** 87–104.

LAWTON, R. (1963) The journey to work in England and Wales: forty years of change, *Tijdschrift voor Economische en Sociale Geografie*, **54,** 61–9.

MEDHURST, F. (1963) Traffic induced by central area functions, *Town Planning Review*, **34** 50–60.

STARKIE, D. N. M. (1967) Traffic and industry: a study of traffic generation and spatial interaction, *L. S. E. Geographical Papers*, **3.**

On models of urban growth see:

COWAN, P., IRELAND, J. and FINE, D. (1967) Approaches to urban model building, *Regional Studies*, **1,** 163–72.

HALL, P. (1967) New Techniques in regional planning: experience of transportation studies, *Regional Studies*, **1,** 17–21.

MORRILL, R. L. (1965) Migration and the spread and growth of urban settlement, *Lund Studies in Geography, series B*, **26.**

THE HUMAN ECOLOGY OF CITIES

A seminal work in this field was:

PARK, R. E., BURGESS, E. W. and MCKENZIE, R. D. (1925) *The City*, U. of Chicago P., Chicago.

See also:

PARK, R. E. (1952) *Human Communities*, Free P., Glencoe.

Various theories of urban structure are discussed in:

GITTUS, E. (1964) The Structure of urban areas, *Town Planning Review*, **35,** 5–20.

HAGGETT, P. (1965) *Locational Analysis in Human Geography*, Arnold, London, especially pp. 177–8.

Studies by geographers of particular cities, using modern techniques include:

CAREY, G. W. (1966) The regional interpretation of Manhattan population and housing patterns through factor analysis, *Geographical Review*, **56,** 551–69.

ROBSON, B. T. (1966) An ecological analysis of the evolution of residential areas in Sunderland, *Urban Studies*, **3,** 120–42.

THOMAS, E. N. (1960) Areal associations between population growth and selected factors in the Chicago urbanized area, *Economic Geography*, **36,** 158–70.

Population gradients are discussed in:

BERRY, B. L. J., SIMMONS, J. W. and TENNANT, R. J. (1963) Urban population densities: structure and change, *Geographical Review*, **53,** 389–405.

CLARK, C. (1951) Urban population densities, *Journal of the Royal Statistical Society, series A*, **114,** 490–6.

NEWLING, B. E. (1964) Urban population densities and intra-urban growth, *Geographical Review*, **54,** 440–2.

The economic background to residential and other urban land values is explored in:

ALONSO, W. (1965) *Location and Land Use: towards a general theory of land rent*, Harvard U.P., Cambridge, Mass.

CITIES AS CENTRAL PLACES

A full translation of Christaller's seminal work is:

CHRISTALLER, W. (1966) *The Central Places of Southern Germany*, translated by C. Baskin, Prentice-Hall, Englewood Cliffs. Originally published in 1933 as *Die zentralen Orte in Süddeutschland*, Fischer, Jena.

For earlier presentation in English see:

DICKINSON, R. E. (1947) *City, Region and Regionalism*, Kegan Paul, London. A revised version of this book, *City and Region: a geographical interpretation*, Routledge, London, was published in 1964.

ULLMAN, E. (1941) A theory of location for cities, *American Journal Sociology*, **46**, 853–64.

For further developments in this field see:

BECKMAN, M. J. (1958) City hierarchies and the distribution of city size, *Economic Development and Cultural Change*, **6**, 243–8.

BERRY, B. J. L. and GARRISON, W. L. (1958) Functional bases of the central place hierarchy, *Economic Geography*, **34**, 145–54.

BUNGE, W. (1962) Theoretical geography, *Lund Studies in Geography*, series C, **1**, revised edn., 1966.

ISARD, W. (1956) *Location and Space Economy: a general theory relating to industrial location, market areas, land use, trade and urban structure*, M.I.T.P., Cambridge, Mass.

LÖSCH, W. (1954) *The Economics of Location*, translated by Woglom, W. H. and Stolper, W. F., Yale U.P., New Haven.

Two examples of empirical studies are:

BRUSH, J. W. (1953) The hierarchy of central places in south-west Wisconsin, *Geographical Review*, **43**, 380–402.

SMAILES, A. E. (1944) The urban hierarchy in England and Wales, *Geography*, **29**, 41–51.

On shopping centres see:

BERRY, B. J. L. (1967) *Geography of Market Centres and Retail Distribution*, Prentice-Hall. Englewood Cliffs.

PROUDFOOT, M. J. (1937) City retail structure, *Economic Geography*, **13**, 425–8.

VANCE, J. E., JR. (1962) Emerging patterns of commercial structure in American cities, in NORBORG, K. (Ed.) *Proceedings of the I.G.U. Symposium on Urban Geography, Lund 1960*, Gleerup, Lund, pp. 485–518.

THE NEED FOR A CULTURAL CONTEXT

A few examples of the rapidly growing literature in this field include:

HORVATH, R. J. (1968) Towns in Ethiopia, *Erdkunde*, **22**, 42–51.

MABOGUNJE, A. (1962) *Yoruba Towns*, University P., Ibadan.

MCGEE, T. G. (1967) *The South-East Asian City: a social geography of the primate cities of south-east Asia*, Bell, London.

SJOBERG, G., (1960) *The Pre-Industrial City: past and present*, Free P., Glencoe.

SKINNER, G. W. (1964) Marketing and social structure in rural China, *Journal of Asian Studies*, **34**, 3–43.

WHEATLEY, P. (1965) "What the greatness of a city is said to be"; reflections on Sjoberg's "Pre-Industrial City", *Pacific Viewpoint*, **4**, 163–88.

PART III

APPLIED GEOGRAPHY

You know my methods. Apply them.
CONAN DOYLE, *The Sign of Four*

CHAPTER 19

GEOGRAPHY AND PHYSICAL PLANNING

D. THOMAS

THE RÔLE OF APPLIED GEOGRAPHY

The present importance of applied geography cannot reasonably be contested. It occupies its fair share of the teaching to undergraduates in British universities, and has become one of the dominating fields of postgraduate study. Interest is equally strong in other parts of the world. Of the fourteen sections under which papers were classified for reading at the International Geographical Congress held in London in 1964 one dealt specifically with applied geography. Only four sections of the programme attracted more contributors.

But it was not always so. As Leszczycki and Wise have shown recently in two papers dealing with the origins and progress of applied geography, the germ of the idea that geographical concepts, techniques, and methods should be applied to the study of the everyday problems of society lies deep in the history of modern geography (Leszczycki, 1964; Wise, 1964). Herbertson, as long ago as the late 1890's, argued that the collection and collation of geographical knowledge could greatly benefit commerce, medicine, missionary work, and the armed forces, but there was little immediate development of such early hopes (Herbertson, 1898). Geography was still a largely descriptive subject with a very modest technical basis, and the possibilities of applying the type of material which it produced were very limited. It was not until geography moved into a more analytical phase in the period between the world wars—the same period in which the subject was becoming established in universities—that more widespread application became feasible.

It was at this time that a single piece of work was undertaken which, more than any other, led to the particular character of British applied geography. In 1930 Stamp began the Land Utilisation Survey, in co-operation with many professional and amateur geographers throughout the country (Stamp, 1948). However, it was not the great body of data and analysis which made the greatest contribution to furthering the application of geography, but the experience and technical expertise which the surveyors and authors

199

had gained in the handling of land-use material. The Second World War presented problems of many kinds which geographers were able to tackle. Specialists in all fields applied themselves to military and naval intelligence, to terrain study, to airphotograph analysis, and to cartographic work. But a number of the land-use specialists quite naturally found themselves engaged in the committees, commissions, and working parties dealing with the re-organization of land use which would take place when the war was over. In 1945, when specialists in other fields were returning to their pre-war pursuits, those connected with land-use planning sometimes became established in the new departments of central and local government charged with the operation of the new planning legislation. Meanwhile they maintained close contact with their colleagues who had returned to more strictly academic work. So it came about that the applied geography which emerged in the years after 1945 was principally concerned with the physical planning of the landscape.

There were also other good reasons why geographers became attracted to physical planning. Not only was it a field in which they had some technical competence, but in the immediate post-war period it was also one in which social needs were great. As Wise has indicated, there was need for the im-provement of the environment to remove the scars of war, there was need to relieve the long-standing housing and landscape problems of the nineteenth-century industrial areas, and there was need to make an orderly readjustment of the geographical distribution of employment and population to meet the changed technical and social conditions (Wise, 1964). To these problems both government and academic geographers applied themselves. New graduates were equally attracted and town and country planning became one of the major career outlets for university-trained geographers, particularly following the Town and Country Planning Act of 1947, which required local planning authorities to undertake surveys of their areas and to produce development plans. As a result, a continuing demand for people who were skilled in survey and map analysis was created, and a second generation of geographers moved into public service in the field of physical planning. It is a movement which has since been maintained.

The relations between geography and physical planning in Britain are deep and manifold. Over a fairly extended period geography has serviced physical planning and, in turn, physical planning has profoundly influenced geo-graphical change. The effects of physical planning upon town and country provide the bulk of the subject matter of applied geography as it is currently taught in British universities. This chapter attempts to assess the geographical contributions to physical planning. The range of these contributions is wide, but broadly they can be thought of as operating at a number of different planes: conceptual, technical, factual and analytical, governmental, and pedagogic.

CONCEPTUAL CONTRIBUTIONS

No branch of human geography has failed to provide ideas useful in physical planning, but clearly it is those aspects of the subject which deal with current problems that have provided the greatest stimulus. Economic geography, particularly in its resource, agricultural, and industrial branches, has been one of the major suppliers of concepts, but it would be difficult to argue that as a generator of ideas it surpasses urban geography, transport geography, social geography, political geography, or population geography, especially as there is considerable overlap among these different sections of human geography. That physical planners have adopted these concepts is hardly surprising. They aim to so reorganize the landscape that it functions more efficiently and at the same time better satisfies generally agreed social goals. To achieve this it is plain that physical planners require a very clear understanding of the existing landscape and of the mechanisms which control it. So many geographical ideas have been absorbed into planning that to produce a list would not only be very difficult, but tedious. Instead, a single body of ideas will be considered which, over the last 15 years, if not over a longer period, has probably influenced the thinking of planners more than any other concept.

Central-place studies, as they have come to be known, deal with the location, size, nature, and spacing of production and market nodes. Since the economic activities of manufacturing and distribution are enmeshed within towns and cities, are closely allied to both primary and tertiary industries, are served by a communication system, and employ large numbers of people who often live at some distance from their places of work, the ideas which these studies produce are vital to planning. The field of study is also extremely interesting from another standpoint. Probably more than in any other branch of human geography, in central-place studies it is possible to observe the convergence of, and indeed to build upon, two quite separate lines of approach. On the one hand there now exists a considerable body of theoretical postulations, based almost entirely upon deductive reasoning and leaning very heavily upon mathematical analysis. On the other, there is an equally rich collection of empirical studies using inductive methods and stemming largely from investigations carried out in the field. As long ago as 1961, when Berry and Pred compiled their most valuable annotated bibliography of central-place studies, the volume of work was already substantial (Berry and Pred, 1961). It has since expanded greatly. The two approaches have been mutually enriching and the attention paid to the work by planners is, in part, a reflection of this.

The outlines of both the theoretical and empirical attitudes to central places have been quite adequately presented elsewhere (e.g. Berry, 1967;

Johnson, 1967, Chapter 5). It remains to examine some of the applications of central-place studies in physical planning. The fact that towns and cities supply goods and services to a reasonably well-defined hinterland, that they, in turn, through the social and economic links which evolve, derive support from that area, including part of their work-force, and that towns and cities can be demonstrated both by theoretical and empirical means to be structured hierarchically, according to the levels of services which they supply, has immediate relevance in the field of regional planning. Large cities, by their interactions with the more rural areas surrounding them, create nodal or functional regions which may form a convenient and logical basis for regional organization. The idea goes back to at least 1919, but it is one which is currently very much alive as new regional systems of physical and economic planning are developing, and as local government reforms are being actively discussed (Fawcett, 1919; Senior, 1966). These problems are discussed further in Chapters 20 and 22. On a much smaller scale the same ideas have been widely employed by local authorities in drawing up their development plans, especially where decisions were called for upon the future development of service centres, and upon the range over which these services should be furnished. For some time such ideas have been included in the training of town planners as a matter of course (see, for example, Keeble, 1964, Chapters 4 and 5).

In other countries planners have also used concepts derived from central-place studies. Berry has recently cited a number of examples (Berry, 1967, Chapter 8). The Dutch, in planning settlements to serve newly established farms upon the Ijsselmeer polderlands, created a two-level hierarchy of market centres. The Israelis, in the recently settled Lakhish plains running eastward from the Gaza strip, adopted a three-level hierarchical system for settlements. In the United States a clear distinction has been drawn when developing shopping plazas between neighbourhood, community, and regional services. To these very practical applications of central-place ideas, Berry added some examples of the use of theory in solving problems. Theoretical frameworks have been used to consider the difficulties of agriculture and rural life in Saskatchewan, to probe the changes needed in the urban structure of Ghana, and in developing models to aid the attack on commercial blight in the Chicago and Toronto regions.

TECHNICAL CONTRIBUTIONS

Despite the very specific research which he now undertakes, there is a tradition, which dies hard, that the geographer is a generalist. It therefore sometimes comes as a surprise that physical planners, who often think of themselves as employing the end-products of other disciplines, consider geographers to be specialists. There is, of course, justification in the planners' view. As has already been argued, geographers first entered physical planning because they possessed particular experience and special expertise in land-use analysis, and since that time methods and techniques within geography have advanced greatly. It is upon this technical competence in the handling of data, as well as upon his rather special conceptional contributions, that the geographer's present reputation in physical planning has been built.

Probably the foremost of his methods to pass into planning are the geographer's mapping and ground-survey techniques. It would be presumptuous to claim that any of these methods are exclusively geographical, or indeed to deny that they have been further modified and refined by geographers and others within the field of planning in order to meet special needs not normally encountered in purely geographical work. But they do constitute a very wide range of techniques which have been used and developed largely by geographers and which are immediately applicable to the problems of physical planning (Stamp, 1960). Jackson, a Birmingham geographer who has since become closely associated with land-use planning, has recently illustrated the diversity of the techniques available (Jackson, 1963). The methods described will naturally be familiar to most geographers. Instead of dealing with them in detail attention will now be directed to one particular method, the use of which has lately increased rapidly, and which geographers, especially, have employed for land-use studies.

Airphotograph interpretation has now become one of the major means of acquiring land-use data. As a method it has a number of distinct advantages over the conventional system of collecting information by ground survey. To begin with, the data are more uniform in quality, since they are normally produced by a few skilled photo-interpreters, rather than by larger numbers of fieldworkers, some of whom must inevitably be less able than others. For the same reason data derived from airphotograph survey are usually found to be more accurate. Certainly the method produces a more coherent body of material, because air surveys are usually completed in a few days, while ground surveys may take months or even years to complete. Perhaps the greatest advantage in this branch of applied geography is the flexibility which airphotography gives. A field survey, whether it is meant to produce a general picture of land use, like the Land Utilisation Survey of Britain, or whether it is designed to provide answers to a particular problem, must be limited in scope. These limitations obviously hinder the use of the data for the solution of other, and perhaps quite different, problems at a later stage. Airphotographs, on the other hand, do not influence problem design in the same way. They can be constantly and swiftly re-interpreted, and indeed, may be made to yield more information as time goes by with improvements in photogrammetric technique.

Although the Second Land Use Survey of Britain did not employ air-photograph analysis, many large-scale surveys carried out in other parts of the world over the last 15 years have used these techniques. For example, Rawson and Sealy (1959) used airphotographs to determine the land-use distribution of Cyprus, and Ward (1965) employed airphotograph analysis in his study of land use and population problems in Fiji. In Sweden, air-photograph maps are being printed upon which land use and other information derived from the photographs is superimposed (Stamp, 1960, p. 141). It must be conceded that airphotograph interpretation is easier for areas which are dominantly rural than for those which are heavily urbanized, but analysis of urban areas is certainly not excluded (e.g. Thomas, 1964). A summary of the use of airphotographs in the interpretation of rural land use in Western Europe in recent years has been provided by Haefner (1967).

Airphotographs have not been nearly as fully exploited in physical planning in Britain as they might have been. Occasionally local planning authorities have used airphotographs in their development-plan survey work, or for other special purposes, but largely it is a neglected technique. At the moment air-survey methods constitute more a potential than an actual technical contribution to land-use planning.

FACTUAL AND ANALYTICAL CONTRIBUTIONS

There is a revulsion in modern geographical circles against data-grubbing for its own sake, and rightly so. Surveys are not ends in themselves. They may be used to teach students techniques, but in more advanced work they are simply a means of answering questions which, without further field investigation, could not have been answered. Hence most geographical work has become what is sometimes described as "problem-oriented". Since human geographers have a deep interest in the problems of cities, towns, and villages, since they are concerned with population distribution, since the location and functioning of economic phenomena are of vital importance, and since the characteristics of the communication system have for long attracted their attention, it follows that much of current geographical fact collecting and analysis is of direct relevance to the physical planner.

There are large numbers of such studies. They range in scale and nature from Hall's polemical *London 2000* (Hall, 1963) to more modest attempts to establish the facts of, and reasons lying behind, some currently contentious situations (e.g. Thomas, 1964). A single example will serve to illustrate studies of this kind. In 1967, with the help of a number of undergraduates from University College London, Perry undertook a study of air traffic to and from Jersey (Perry, 1968). The work was based upon an examination of airport records, of national statistics on passenger and freight movement by air, and upon a field survey of hotels and boarding houses in the two principal holiday centres of Jersey, St. Helier and St. Aubin. The study revealed the importance of the airport and air transport to Jersey and also the effects which both had upon the landscape of the island.

The airport itself engaged 4·4 per cent of the total island labour force, and because many of the workers were highly skilled, this had important economic effects as well as generating a demand for houses in a small and tightly-packed island. Passenger traffic into the airport was increasing rapidly. Some of the increase was contributed by an expansion in tourism, but part was due simply to air travel becoming the usual means of entering Jersey, at the expense of sea transport. The bulk of the passenger traffic originated in south and southeast England; but since the hotel and boarding-house survey showed that a high proportion of holiday-makers lived in the north of England, there was a clear indication that many holiday makers travelled south-wards by land and then made a short air-crossing. They were thus able to cross the water comfortably and speedily, and yet minimize total journey cost. Airfreight traffic had expanded even more rapidly than passenger traffic. Roughly 10 per cent of total imports and exports, by weight and value, were carried by air, the chief imports being newspapers, consumer goods, and foodstuffs, while the chief exports were market-garden produce, particularly flowers, and fabrics and textiles. About 80 per cent of Jersey's trade is with the United Kingdom, to which it has free-trade access.

The implications for land-use development of further increases in air transport are clear. The island measures 9 by 5 miles and small external changes can have a big impact. But, conversely, so may internal and external changes affect air transport. One of the most profound changes that might possibly take place is the entry of Britain into the European Economic Community. If Jersey entered too, then it would face stiff competition from the market gardeners of Italy and the Netherlands, and it might also be compelled to harmonize its taxation and financial policies with those of other members. If it did not enter, then it would face Community tariffs. Either way, air passenger and freight traffic would be strongly influenced.

CONTRIBUTIONS TO GOVERNMENTAL WORK

The beginnings of the links between geography and government in England and Wales have already been outlined in the early part of this chapter. It is now necessary to specify some of the particular fields of governmental physical planning to which geographers have contributed.

Of the many geographers who became involved in planning during the 1939–45 war as research, regional planning, and maps officers, a number eventually rose to positions of authority and influence. For example, E. C.

Willatts, who had been closely associated with L. D. Stamp in the Land Utilisation Survey, became Maps Officer of the Ministry of Housing and Local Government. He was responsible for many of the maps in the 1:625,000 series and also for the desk atlas, produced by the Ministry for its own use, and for the assistance of local planning offices. These maps later provided the basis for some of those appearing in the *Atlas of Britain*, published in 1963. J. R. James became chief technical planning officer in the same Ministry before moving back to purely academic work in 1967, and A. G. Powell is effectively in charge of the Ministry's study group on southeast England, following up the work already undertaken by the South-East Economic Planning Council and the South-East Conference of Local Planning Authorities, and attempting to produce a regional plan for the area. In addition to these men the Ministry of Housing and Local Government alone employs 40 geography graduates, local authorities employ at least 200 more, and rapidly increasing numbers are engaged in the work of the economic planning boards.

At a more advisory level the contributions of geographers have also been great. For example, Stamp himself was vice-chairman of the Scott Committee on Land Utilisation in Rural Areas, which reported in 1942, S. H. Beaver and S. W. Wooldridge were members of the Advisory Committee on Sand and Gravel, which reported in 1948, three academic geographers at one time served concurrently as National Parks Commissioners, and geography is represented on such bodies as the Nature Conservancy and the Water Resources Board, and upon committees inquiring into the problems of agricultural small-holdings and allotments. These are formal, public duties. Geographers also act more privately as consultants to government departments and local authorities. Together these activities represent a greater participation in governmental action than has ever existed previously.

PEDAGOGIC CONTRIBUTIONS

At the moment, the major springboard for geographers to careers in physical planning is undoubtedly provided by the many courses in British universities which consider the geography of planning. Within departments of geography these masquerade under a number of different disguises. Occasionally they are entitled "geography of planning", more often "applied geography", but frequently the subject of physical planning also forms a substantial part of courses in economic, urban, and settlement geography, as well as featuring in courses on regional geography. At the postgraduate level many specialized courses exist. Some of these are wholly within geography departments, but increasingly they rely upon interdisciplinary teaching in which the geographical, social, administrative, and economic aspects of physical planning are considered. In some universities even more specialized courses in such subjects as conservation are offered.

But these courses are usually non-professional in the sense that, though they give general training in the field of physical planning, they do not lead to, nor exempt candidates from professionally recognized examinations, such as those of the Town Planning Institute. Such courses are given only within academic departments concerned solely with town and country planning. However, even here geographers contribute. An increasing number of these departments have geographers on their staffs, and many more find it necessary to invite geographers to give special courses to their students. As long ago as 1958 T. W. Freeman wrote a book setting out the geographical background to planning, which was designed to support just such a course (Freeman, 1958).

CONCLUSION

The foregoing chapter is inevitably partial in its view. It has dealt with one aspect of applied human geography, largely with reference to Britain. It has ignored much of the applied geographical work under way in other countries, for example France (Phlipponneau, 1960), Belgium (Comité National de Géographie, 1964), or the United States (e.g. Borchert, 1961). It has ignored the whole field of applied physical geography, touched upon elsewhere in this volume (e.g. Chapters 5 and 6). However, it has outlined one field in which geographers have contributed substantially to thought, technique, data, analysis, administration, and teaching, and in which, at the same time, they have arrived at conclusions useful to their fellow-men. While usefulness, in itself, can never be the sole criterion by which to judge scholarly study, one cannot but agree with Linton that "geographers will surely not be happy in their work unless they can see some relation in its purpose to the current goals of human endeavour, and can relate its practice in some way to the needs of the times" (Linton, 1957).

REFERENCES AND FURTHER READING

THE RÔLE OF APPLIED GEOGRAPHY

The general field of applied geography is considered by:

LESZCZYCKI, S. (1964) Applied geography or practical applications of geographical research, *Geographica Polonica*, **2,** 11–21.
WISE, M. J. (1964) The scope and aims of applied geography in Great Britain, *Geographica Polonica*, **2,** 23–36.

One of the early accounts of the uses of geography is to be found in:

HERBERTSON, A. J. (1898) Report on the teaching of applied geography, *Journal of the Manchester Geographical Society*, **14,** 264–85.

An account of the progress and results of the Land Utilisation Survey is contained in:

STAMP, L. D. (1948) *The Land of Britain, its Use and Misuse*, Longmans, London.

CONCEPTUAL CONTRIBUTIONS

General accounts and further bibliographies of central-place studies are to be found in:

BERRY, B. J. L. and PRED, A. (1961) *Central Place Studies: bibliography and review*, Regional Science Research Institute, Philadelphia.

BERRY, B. J. L. (1967) *Geography of Market Centers and Retail Distribution*, Prentice-Hall, Englewood Cliffs.

JOHNSON, J. H. (1967) *Urban Geography: an introductory analysis*, Pergamon, Oxford.

Examples of older and newer applications of central-place studies are contained in:

FAWCETT, C. B. (1919) *Provinces of England*, reprinted by Hutchinson, London, 1961.

KEEBLE, L. (1964) *Principles and Practice of Town and Country Planning*, Estates Gazette, London.

SENIOR, D. (Ed.) (1966) *The Regional City*, Longmans, London.

TECHNICAL CONTRIBUTIONS

Geographical techniques used in practice are described by:

JACKSON, J. N. (1963) *Surveys for Town and Country Planning*, Hutchinson, London.

STAMP, L. D. (1960) *Applied Geography*, Penguin, Harmondsworth.

Examples of the use of airphotographs are contained in:

HAEFNER, H. (1967) Airphoto interpretation of rural land use in Western Europe, *Photogrammetria*, **22,** 143–52.

RAWSON, R. R. and SEALY, K. R. (1959) *Land Utilization Map of Cyprus*, Geographical Publications, Bude.

THOMAS, D. (1963) London's green belt: the evolution of an idea, *Geographical Journal*, **129,** 14–24.

WARD, R. G. (1965) *Land Use and Population in Fiji*, H.M.S.O., London.

FACTUAL AND ANALYTICAL CONTRIBUTIONS

Of the many studies under this head the following may be cited as examples:

APPLEBAUM, W. and COHEN, S. B. (1960) Evaluating store sites and determining store rents, *Economic Geography*, **36,** 1–35.

HALL, P. (1963) *London 2000*, Faber, London.

PERRY, N. H. (1968) The air traffic of Jersey (Channel Islands), *Tijdschrift voor Economische en Sociale Geografie*, **59,** 159–64.

THOMAS, D. (1964) The components of London's green belt, *Journal of the Town Planning Institute*, **50,** 434–9.

CONTRIBUTIONS TO GOVERNMENTAL WORK

For a short account see:

STEEL, R. W. (1963) British geographers, in WATSON, J. W. and SISSONS, J. B. (Eds.) *The British Isles: a systematic geography*, Nelson, London, pp. 20–39.

PEDAGOGIC CONTRIBUTIONS

A book providing students of planning with a geographical background is:
FREEMAN, T. W. (1958) *Geography and Planning*, Hutchinson, London.

CONCLUSION

Examples of applied geographical work overseas are found in:

BORCHERT, J. R. (1961) The Twin Cities urbanized area: past, present, future, *Geographical Review*, **51**, 47–70.

Comité National de Géographie (1964) *Les Applications de la Géographie en Belgique*, Secretariat du Comité National de Géographie, Liège.

LESZCZYCKI, S. (1960) The application of geography in Poland, *Geographical Journal*, **126**, 418–26.

PHLIPPONNEAU, M. (1960) *Géographie et Action*, Colin, Paris.

The rôle of geography in a changing world is considered by:

LINTON, D. L. (1957) Geography and the social revolution. *Geography*, **42**, 13–24.

CHAPTER 20

GEOGRAPHY AND ECONOMIC PLANNING

P. A. WOOD

THIS chapter is concerned with the spatial aspects of economic activity and the attempts of public authorities to plan them as part of the wider task of promoting economic development. Four themes will be discussed in order to examine the possible contribution of geography to this important aspect of modern economic organization. (1) The modern significance of planning, from the local to the international scale. (2) The significance of spatial variations in economic planning and the manner in which geography as a discipline is facing the challenge of planning requirements. (3) How can our current view of complex spatial relationships be used in planning at different stages of economic development and how can this understanding be incorporated into the teaching of the subject? (4) By way of exemplification, the rôle of regional definition as a device to aid the planning of the spatial economy will be examined in some detail.

THE RÔLE OF PLANNING IN MODERN SOCIETY

Planning has become part of our lives. The reasons for this are related to the growing size and complexity of modern society. Before the Second World War reference to the institutions, power, and theory of planning implied one of two realms of action. The first, at the *national* level, comprised policies aiming to achieve desirable economic characteristics, such as growth, stable employment and prices, the equitable sharing of wealth, and a satisfactory balance of payments. Of course, forms of national economic policy had been established well before the Industrial Revolution, but after about 1870 governments began to increase their powers to control the changes induced by growing trade and industrialization. Since then, "the change in the relation between governments and economic life . . . [has been] one of the most fundamental of all changes in social organization in the period" (Ashworth, 1962, p. 160).

The second realm of planned action was concerned with essentially *local* town planning measures for the control and promotion of physical building

developments (Ashworth, 1954; Cullingworth, 1967). Like national economic policies, this grew rapidly during the last quarter of the nineteenth century, when the squalor of the industrial cities became too great to remain
socially and politically tolerable and it was recognized that individual
decisions could have harmful social consequences.

The methods of national economic planning and the extent of local physical
controls have come to vary widely in the twentieth century among countries
at differing stages of development and with different systems of government.
It has also become evident that planning at these two levels has not effectively
dealt with the complexity of the world with which it is concerned. This has
been an important consideration in the more modern growth of two other
areas of planned action.

One of these originated between the wars, when programmes were implemented to develop land, water, and mineral resources within specific *regions*
in countries as diverse in their political philosophies as the United States and
the Soviet Union (Friedman and Alonso, 1964, part IV). Another stimulus
to regional planning at that time came from unemployment which drew
attention to the impact of economic obsolescence, localized as it was into
certain settled and formerly prosperous regions. Policy makers thus began to
recognize the significance of economic variations within national boundaries
and to consider measures that would alleviate their most harmful effects.
Regional economic planning took on a comprehensive, institutional form in
many countries only in the 1950's, however, when time and greater prosperity
had shown that the mere adjustment of national policies could not produce
major changes in regional differences of economic growth. By then also the
large-scale problem of metropolitan cities had intensified to demand co-
ordinated public action at regional levels (Hall, 1966).

A further set of problems requiring planned action based upon an understanding of their causes are *internationl* in scale. Economic planning at the
international level is still poorly developed but it is now widely recognized
that the plight of poor countries is inextricably associated with the prosperity
of the rich. The familiar and pressing supra-national issues of trade and
growth, free-trade areas and common markets, international commodity
agreements and overseas aid, well illustrate the general difficulty of transforming theoretical economic interpretations into effective policies at all
levels of planning.

In broad terms, therefore, planning has arisen from needs that have
become implicit in society at particular stages of development. The complexity of social and political problems has progressively extended the areas of
life that men attempt consciously to mould, and planning effort in one form
or another now influences virtually every aspect of human organization.

THE SIGNIFICANCE OF
PLANNING FOR GEOGRAPHICAL STUDIES

At each of these scales—national, local, regional, and international—the key economic issues revolve round disparities of wealth and growth between places. The crucial importance of spatial or geographical considerations should thus be clear (Ullman, 1960): the solution of economic problems is not simply a "local" issue, dependent upon a narrowly defined inventory of physical, biological, and human resources. It is now widely recognized that part of the essence of development problems lies in the position of different points on the earth's surface, affecting in a dynamic way supplies of physical resources, labour and capital, access to transportation routes, and markets and competition from other places, whether they be nations, regions, town or country.

Thus, the settlement, the region, and the developing nation should have brought geographers increasingly into contact with the problems of planning during the last two decades, in common with the other social sciences. Yet, in an essay summarizing recent work in the study of development problems, Keeble is critical of the approach of geographers to these issues in the past (Keeble, 1967). Even at the local and regional scale, the traditional focus of geographical research work, their empirical and descriptive approach to the spatial aspects of economic growth has left, as Keeble puts it, a wide intellectual void that has come to be filled by economists. The major advances in our understanding of regional and local relationships in the 1950's were made by American economists (associated with other workers, including a few geographers) who developed the field of "regional science", encompassing such diverse disciplines as economics, sociology, demography, geography, and history (Isard, 1960). More recently, a very valuable set of essays on post-war developments in economic theory included an extensive survey and bibliography of "regional economics", indicating how regional and local studies are rapidly being integrated into the larger body of economic thought (Meyer, 1967).

The nature of economic planning therefore places a strong emphasis upon spatial problems and has prompted a search for suitable analytical techniques to be used in their investigation. Economists have led the way but the theoretical and analytical bias of much recent work in geography might be expected to offer many new insights, and some hope for major geographical contributions in the future. It is important that the significance and value of this new approach to the subject should become generally accepted at all levels of teaching.

All disciplines that wish to contribute towards the effectiveness of both physical and economic planning find severe demands placed upon them. Not least among these are the problems of data collection and the development of computer storage and processing techniques. More profoundly, however, planning studies require a willingness to specify and test ideas and generalizations—to reinforce our supposed understanding of society with precision and even, ultimately perhaps, with accurate predictions. Modern geographical studies into the spatial patterns of transportation, land use, urban growth, industrial location, or service facilities are attempting to strengthen the logic and accuracy of our modes of thought at least partly for these purposes; the rôle of apparently abstract theory may be essentially practical, in a way that "empirical" geography has never been in the past. This is the purpose of experiment, the setting up and verification of hypotheses about cause and effect, and, above all, defining clearly what we can and cannot be certain about. It is only as an aid to the process that "quantification"—the use of mathematical notation and statistical techniques—has any significance. Current academic controversy in geography revolves around these questions and the process has been stimulated very largely by the need for answers to questions of planning policy, with a resulting awareness of the limitations of our understanding of spatial processes.

THE SPATIAL DIMENSION IN ECONOMIC PLANNING

The value of any contribution that geographers may make to economic planning in the future depends particularly, therefore, upon the rigour and precision with which they deal with spatial variations, relationships, and linkages. As Friedmann and Alonso remarked (Friedmann and Alonso, 1964, p. 1). "The social sciences, principally economics and sociology, have been laggard in taking notice of space; while geography, which has always dealt with space, has lacked analytic power."

This "analytic power" must grow from an appreciation of the general nature of the relationships with which geography is concerned throughout the teaching of the subject. Gaining such an appreciation of spatial relationships does not, of course, necessarily involve the more complex techniques of analysis devised for regional science. Indeed, Meyer remarks that regional economists "have shown a great propensity for undertaking the difficult tasks first" (Meyer, 1967, p. 265) and suggests that there is need simply for the preliminary testing of ideas about such problems as the behaviour of firms and social groups and the sources and effects of investment at the regional level. In this vein, research geographers are now concerned with testing some of the fundamental ideas about spatial relationships that may be of value in economic planning through their analyses of the effects of distance upon the strength of various types of linkage, the characteristics of networks, the accurate description and analysis of patterns of settlement, the definition of regions, and the diffusion of innovations (Haggett, 1965; Chorley and Haggett, 1967; Yeates, 1968; Berry and Marble, 1968).

Perhaps, as a framework into which these studies might be fitted to illustrate their significance for economic planning, the term "economic surface" suggests the aggregate effect of the whole group of relationships between function and spatial patterns that they examine. Modern studies in human geography are concerned with what is best regarded as a continuum of relationships and connections from place to place, whose nodes result in peaks of human activity and exchange, linking networks of movement and aggregate "surfaces", representing varying intensities of human economic activity over the earth's face. The height of such surfaces at any point or for any areal unit might be measured by the value of economic activity generated there over any period of time. The generalized topography of these surfaces changes between types of economy; with low areas and dangerous peaks and precipices characterizing underdeveloped countries, and plateaux and gentler gradients in more advanced economies. Thus, different levels of economic development are associated with particular types of spatial and regional problem, and perhaps in this we may see how the different realms of planning action outlined earlier; international, national, regional, and local, are interconnected.

STAGES OF ECONOMIC AND SPATIAL DEVELOPMENT

Broadly, Friedmann has suggested (Friedmann, 1966) that the organization of economic linkages between places is generally of subsidiary importance in pre-industrial economies, with less than 10 per cent of the Gross National Product (G.N.P.) derived from manufacturing. The first priority of planning must be given to achieving economic "take-off" (Rostow, 1960 and 1963; Keeble, 1967) by establishing industries and the necessary infrastructure of educational, health and agricultural organization and transportation facilities. During the transitional phase to an industrial economy, when manufacturing builds up to 25 per cent of the G.N.P., rapid growth, combined with low spatial mobility, normally has an obvious and acute geographical reflection in the concentration of activities into one or a few centres. These have been termed "core regions" or "growth poles", and form nodes of concentrated, interlinked economic activity in manufacturing, services, and commerce, which exert strong influence upon the economic development of the whole country. Characteristically, in countries such as Venezuela, Brazil, Mexico, Turkey, India, or Pakistan, this influence threatens the full use of resources over their whole area; the "core regions" have their own problems of rapid urbanization and also drain the dynamic element from other regions. The common result is rural depopulation to urban areas that cannot cope with the influx, and this geographical aspect of development has been termed the "centre–periphery dichotomy". This dichotomy has been traced on global, continental, national, and sub-national scales (Friedmann, 1966, pp. 10–13), and appears to be the spatial equivalent of and complement to the time scale of Rostow's "take-off" sequence. As a generalization about the form of the spatial economy it has similar uses and limitations, being based essentially upon a rationalization of historical experience. It exists in modified and more complex forms even in developed countries and has become probably the major geographical theme of economic development and planning. It has also been incorporated into the highly influential general theories of economic growth of Myrdal and Hirschmann (Myrdal, 1957; Hirschmann, 1958), who analysed the cumulative process by which the dichotomy becomes established at both international and interregional levels.

In mature industrial nations, such as those of western Europe, problems of regional imbalance result from the inevitable adjustments that are needed as new industries rise and old ones decline. No region or place can escape this process but areal specialization results in marked relative decline in certain regions. Obsolescence of old cities and old industries creates a permanent problem of renewal even though all regions may be fully integrated into the national economy (Boudeville, 1966). Thus the problems of growth attracting further growth are not confined to developing countries alone and most developed nations have had to adopt policies aiming to achieve some specified balance between regions. Although the centre–periphery problems are less acute than in developing countries, they have great social and political significance, attracting favourable treatment to rural areas and for rebuilding the environments and economies of old industrial and mining conurbations.

If American experience has general validity, the ultimate stage of economic development is marked by a decline in the contribution of manufacturing to the G.N.P., with services of various types becoming dominant. Geographically the metropolis assumes prime importance and planning becomes preoccupied with urban and metropolitan problems (Gottman, 1961). At this stage the whole spatial economy has become tied in with the functioning of metropolitan centres. The problems of wealthy urban societies, such as the improvement of the quality of environment, urban renewal, the adjustment of land-use competition to the pressures of growth, leisure, and recreational provision, and the achievement of efficiency of communication in relation to the organization of community life are the main planning concerns at this most highly developed level.

THE REGION AND PLANNING

At each of these levels of economic development, with their attendant problems, the accepted device for dealing with spatial planning is the region. Whatever progress is made in our understanding of the apparent continuum of human spatial relationships, there may well remain the necessity in the future to separate and classify sections of the aggregate surface of economic activity for administrative and planning purposes. Will the grouping together of adjacent places into regions be the best way to do this? Regionalization is a well-established aspect of geographical work that has gained much stimulus from practical planning needs. It is perhaps ironic, as regional planning becomes more widely adopted, that the contribution of regional subdivisions to our understanding of the spatial system and their relevance to policy should be questioned.

Clearly, the "core regions" already referred to, have some functional significance, as do the complementary "frontier regions", "development axes", and "depressed regions" suggested by Friedmann and Alonso. If, however, we take a more general approach to spatial organization, the identification of types presents a more formidable task, even at one level of economy. Various levels of development, as we have seen, are associated with different degrees of interlinkage and categories of spatial problem.

There is a long history of controversy in geography about the nature of regions (Grigg, 1967), although this has not prevented planners and workers in related fields from ignoring or assuming away the important conceptual difficulties of regional classification. Very broadly, as this classification has become dominated by urban forms and nodes, the static view of "homogeneous regions", grouping places together because of their similar formal properties of physical geography or agricultural land use, has been replaced by the "nodal region" or "functional region". This groups places together with common linkages and relationships, even if they may be otherwise dissimilar, (e.g. rural areas are related to functionally dominant urban areas).

Today, the approach to regionalization is made, like other aspects of social or biological classification, within a broad "systems" framework (Grigg, 1967; Haggett, 1965, p. 17). In its most general form, systems theory simply suggests that we should look at the world in search of objects (in this case, places) in groups, associated together in some positive manner. Systems are therefore arbitrarily demarcated sections of reality, marked by some common functional connections. Any identifiable nodal region, with its structure of internal linkages separating it by definition from any other, is a system. If a region has no internal structure and no clear principle by which it is separated from other areas, there is little to be gained, in terms of reliable pointers to planning policy, from dealing with such a vacuous entity. All such systems are, however, affected in practice by some general outside influences. In Britain, for instance, we may define and classify regional structures in terms of such special aspects of linkage as journeys to work, travel to large shopping centres, numbers of telephone calls, broad land value patterns, freight movements, kinship ties, or industrial linkage. Any of these classifications may be of value for planning purposes but, of course, each region so defined will be strongly under the influence of national economic, technological, and social trends.

Recalling the general nature of spatial relationships, referred to earlier in the shorthand form of "economic surface", two major reservations might be placed upon the indiscriminate use of regions for dealing with the spatial aspects of economic planning. The first is the obvious one that no single definition can be expected to serve all purposes, reflecting all of the spatial linkages that may be the subject of planned action. As a result, few regional divisions that are in operation fully reflect any of these linkages; they are a compromise. Often in practice the theoretical difficulties are avoided by establishing policy or administrative regions, defined for the purposes of particular development programmes or forms of administration (Acton Society Trust, 1964–5). The value of such regions depends upon the quality of the development programme or the appropriateness of the particular form of administration to deal with current planning problems. The danger of this type of solution is that crudeness of areal classification, particularly in relation to the needs of different localities within the defined regions, may be reflected by insensitive policies, operating as if the regions were internally uniform. Post-war regional policies in Britain have particularly suffered from these hazards and the current (1968) planning regions offer little improvement. This has not been because the numbers, size, or shape of the regional units are necessarily bad, but because they neither have functional validity nor reflect a properly worked out system of regional development or administration.

Second, there is always the problem, elaborated by Boudeville in his analysis of French regional planning policies (Boudeville, 1966, Chapter 1), that other classes of economic space exist than the region, which must consist of groups of places that are contiguous, or next to each other. The factories owned by a single firm, government agencies and nationalized industries, classes of area with high *per capita* incomes, urban areas within a country or areas under a particular type of farming have strong functional similarities that are important for planning, without necessarily being capable of geographical grouping together. To look at it in another way, geographical proximity is not necessarily a sign of the strongest functional linkages.

The derivation of regions to aid the implementation of planning is therefore a difficult problem; there are many aspects of the spatial economy that probably cannot be understood or planned for adequately within the straight-jacket of formal regional subdivisions. It may well be that we should not try to subdivide what we have seen to be essentially a continuum. Perhaps, on the other hand, the problem is simply that regional definitions are taken too seriously once established. Regions have no individual entity or status, at least in economic terms, being simply the convenient results of a classification of the aggregate spatial system. Changing priorities in planning will produce changing regional definitions, each one valid so long as it aids our under-standing of geographical relationships. In a country such as Britain, the reality of this situation may be administratively inconvenient at present, but this is very largely because of the absence of suitably detailed spatial informa-tion (e.g. for small grid squares or even, ideally, for individual points), which would allow a flexible definition of regions at various scales for different policies. We have tended to accept a rigid division of the country into monolithic regions because they are the smallest units for the collection of detailed economic data. This vicious circle, a major obstacle to the success of economic planning within the country, may eventually be broken if geo-graphers adopt a positive rôle in providing planners with accurate and rigorous accounts of the true nature of the spatial economy.

CONCLUSION

The problems of subdividing areas into regions is a good example of how imperfect understanding or consideration of the nature of the spatial economy may cause stereotyped views to become enshrined in planning folklore. How can a training in geography equip students with the basic understanding that will enable them to challenge such assumptions? Unfortunately, the assumption that regions consist of unchanging, homogeneous entities per-vades many textbooks of regional geography. Typically, the regional sub-division of a country is derived from its physical geography and rural life, but as we have seen, the problems of countries of all kinds are increasingly concerned with dynamic economic and geographical relationships. If geo-graphical study at any level is to be concerned with the current problems of the world—if it is to be analytical rather than descriptive—regional classifica-tion should be couched in terms of these relationships. In fact, it might best be left until after the study of the systematic aspects of a country's geography, when some understanding and knowledge can form the basis of a scientific classification. Perhaps, as a practical suggestion, much more critical appraisal of regional textbooks should be encouraged in this light. The key question might then be: how can their facts be organized and augmented to provide a realistic framework for the study of current planning problems?

In the case of Britain, economic planning, including some element of regional planning, is with us. Economic and industrial problems are best dealt with systematically, but the early use of the various official regional studies and plans for examining selected regions seems to be an obvious step towards greater relevance to current planning problems (see References). Again, critical appraisal of these studies in the light of a systematic understanding of the spatial economy is essential, since none of them is ideal, if only because of the basis for subdividing the country upon which they are founded.

Finally, the development of a distinctive understanding of spatial relationships in geographical training can start at the simplest and smallest scale. Experiments might be carried out to illustrate the simple truths of geographical analysis. Using local information such as transport time-tables, census journey to work tables, traffic counts, or the examination of places that are included in the daily lives of individuals and groups at school, college, work, and home, general patterns can be sought through the testing of basic hypotheses. The simplest of these, although not the only one, is that the volume of connections to a place is inversely proportional to its distance away, with some weighting for the relative attractiveness of different places (further ideas might arise from Yeates, 1968). More advanced elaboration of this simple approach might follow through the examination of the relative impact upon individual places of other points at varying distances away. Such analysis should aim to differentiate among the local, regional, national, and international influences that are fundamental to an understanding of economic activity at any place on the earth's surface and are, therefore, also basic for the solution of its economic planning problems.

A better grounding in the fundamental principles of spatial relationships is urgently needed as part of geographical training and planning problems supply the highest relevance to this aim. Even if the simple generalizations do not work perfectly in reality, the reasons for this and their elaboration are the very stuff of modern geography.

REFERENCES AND FURTHER READING

THE RÔLE OF PLANNING IN MODERN SOCIETY

Ashworth, W. (1962) *A Short History of the International Economy since 1850*, 2nd edn., Longmans, London.

Ashworth, W. (1954) *The Genesis of Modern British Town Planning*, Longmans, London.
Cullingworth, B. (1967) *Town and Country Planning in England and Wales*, 2nd edn., Allen & Unwin, London.

For an up-to-date analysis of economic planning including regional planning in Europe see:

Denton, G., Forsyth, M. and Maclennan, M. (1968) *Economic Planning and Policies in Britain, France and Germany*, P.E.P., Allen & Unwin, London.

The literature on problems of international economic development is vast. Among the accessible works that deal with problems of trade and the international economy in relation to underdevelopment are:

BHAGWATI, J. (1966) *The Economics of Underdeveloped Countries*, Weidenfeld & Nicholson, London.

HIRSCHMANN, A. O. (1958) *The Strategy of Economic Development*, Yale U.P., New Haven.

JOHNSON, H. G. (1967) *Economic Policies Towards Less Developed Countries*, Allen & Unwin, London.

KUZNETS, S. (1966) *Modern Economic Growth: rate, structure and spread*, Yale U.P., New Haven.

MYDRAL, G. M. (1957) *Economic Theory and Underdeveloped Regions*, Methuen, London.

NURKSE, R. (1961) *Patterns of Trade and Development*, Oxford U.P., Oxford.

TINBERGEN, J. (1962) *Shaping the World Economy*, 20th Century Fund, New York.

The geographical background is found in:

THOMAN, R. S. and CONKLING, E. C. (1967) *Geography of International Trade*, Prentice-Hall, Englewood Cliffs.

THE SIGNIFICANCE OF PLANNING FOR GEOGRAPHICAL STUDIES

KEEBLE, D. E. (1967) Models of economic development, in CHORLEY, R. J. and HAGGETT, P. (Eds.) *Models in Geography*, Methuen, London, pp. 243–302.

ULLMAN, E. L. (1960) Geographic theory and underdeveloped areas, in GINSBURG, N. (Ed.) Essays on Geography and Economic Development, *University of Chicago Department of Geography Research Paper*, **62,** 26–32.

The economic approach to regions is summarized in:

ISARD, W. (1960) *Methods of Regional Analysis*, M.I.T. P., Cambridge.

MEYER, J. R. (1967) Regional economics, in *Surveys of Economic Theory*, American Economics Association and the Royal Economics Society, **2,** pp. 240–71, and also in NEEDLEMAN, L. (1968) *Regional Analysis*, Penguin, Harmondsworth, pp. 19–60.

THE SPATIAL DIMENSION IN ECONOMIC PLANNING

FRIEDMAN, J. and ALONSO, W. (1964) *Regional Development and Planning: a reader*, M.I.T. P., Cambridge.

The most coherent and accessible presentation of the various aspects of modern spatial analysis is:

HAGGETT, P. (1965) *Locational Analysis in Human Geography*, Arnold, London.

See also:

CHORLEY, R. J. and HAGGETT, P. (Eds.) (1967) *Models in Geography*, Methuen, London, especially part IV, pp. 459–668.

At a simpler level, some of the approaches of modern geography are explained in:

YEATES, M. H. (1968) *An Introduction to Quantitative Analysis in Economic Geography*, McGraw-Hill, New York.

Specific examples of more advanced work are in:

BERRY, B. J. L. and MARBLE, D. F. (Eds.) (1968) *Spatial Analysis: a reader in statistical geography*, Prentice-Hall, Englewood Cliffs.

FRIEDMANN, J. (1966) *Regional Development Policy: a case study of Venezuela*, M.I.T. P., Cambridge.

The theory of economic take-off into sustained growth is explained in:

ROSTOW, W. W. (1960) *The Stages of Economic Growth: a non-communist manifesto*, Cambridge U.P., Cambridge.

Rostow, W. W. (Ed.) (1963) *The Economics of Take-off into Sustained Growth*, Macmillan/St. Martin's P., London.

The classic study of modern metropolitan trends in the U.S.A. is:

Gottmann, J. (1961) *Megalopolis: the urbanized northeastern seaboard of the United States*, M.I.T. P., Cambridge, Mass.

THE REGION AND PLANNING

For a useful bibliography of recent discussion on regional planning, see:

Town and Country Planning, **36**, 301–4 (June 1968).

A valuable account of the planning problems in major world metropolitan areas is:

Hall, P. (1966) *The World Cities*, Weidenfeld & Nicholson, London.

An account of regional planning that combines thoughtful theoretical discussion with practical experience in France is:

Boudeville, J. R. (1966) *Problems of Regional Economic Planning*, Edinburgh U.P., Edinburgh.

A summary of the logical approach to regional definition is in:

Grigg, E. (1967) Regions, models and classes, in Chorley, R. J. and Haggett, P. (Eds.) *Models in Geography*, Methuen, London, pp. 461–510.

Detailed exemplification of the evolution of regional thinking and institutions can be found in a series of pamphlets:

The Acton Society Trust (1964–65) *Regionalism in England*, 3 vols., London.

Regional studies from the Ministry of Housing and Local Government (for the Southeast and West Midlands) and the Department of Economic Affairs had been published to cover all parts of England by late 1968. Certain of the Regional Economic Planning Councils had also published reports. They are available from Her Majesty's Stationery Office, London.

CHAPTER 21

PLANNING STUDIES IN RURAL AREAS

H. D. CLOUT

INTRODUCTION

Most planning studies undertaken in the past have been concerned with urban or general economic problems. Planning for rural areas has long been a Cinderella-like branch of the subject but it is now developing rapidly in response to three major trends. First, modern, largely urbanized societies are making increasing demands on their rural and suburban surroundings and are changing the sociological, economic, and land-use components of these environments which were established in the past (Frankenberg, 1966; Pahl, 1966). It is essential for our general well-being that the modifications which will take place in rural areas should be guided rationally (Arvill, 1967). Second, the technological changes affecting modern farming demand that the scale of agricultural structures (fields, farms, and land-use patterns) should be enlarged for economic viability in the future (Weller, 1967). This is perhaps a more urgent problem in Continental Europe than in much of Britain. Third, many rural areas are rapidly being depopulated and, in view of these changes in occupance, new policies need to be developed for the provision of services in the future and for the expansion or contraction of settlements.

Country planning merits our attention for three main reasons. The problems are "real" ones and are relevant to our students' personal experience of the countryside; the factual information presented in planning reports will keep our teaching material up to date; and a consideration of country planning may provide a more satisfactory means of discussing rural areas and their problems, which tend to be either ignored or confined to a brief mention in the descriptive catalogues of so much of traditional regional geography.

Many studies in country planning lack the stamp of a well-defined discipline of analysis and action. Rather they reflect specialist viewpoints from which a wide range of structural problems have been considered. As a result, an adequate synthesis of theory and practice in country planning has yet to be written. Most information has to be sought from planning reports and

222

articles in learned journals. These studies betray the administrative, economic, geographical, or sociological preoccupations of their authors. All too
often the detailed realities of country planning are hidden in reports which
are not available to the public.

The general objectives of country planning are the improvement of present
structural conditions and the anticipation of problems which may arise in the
future. These practical aspects will be illustrated in the selection of trends
discussed in this chapter. One "theoretical" development must be stressed,
namely the realization that rural problems can no longer be considered only
in the context of the countryside: they must be seen in the light of external
as well as internal forces of change (Fig. 20). An early outline of country
planning puts the interests of the countryman first and the authors warned
that "any attempt to reconstruct the countryside will fail if it does not take
into full account all the circumstances of his life" (Orwin, 1944, p. 1). In the
quarter century since those words were written, increased personal mobility
and the diffusion of mass media of communication have brought the occu-
pants of many rural areas into a broader, "mentally urbanized" way of life
(Pahl, 1965, p. 5). The demands of some city-dwellers for houses in the country and for increased rural recreational facilities have blurred some of the
distinctions between the more remote countryside, with its contracting
agriculture and ageing population, and "rural" areas within quick and easy
access to urban areas. As a result of these changes, the aims of modern
country planning have broadened to form part of regional physical planning.
This point is explored fully by both Arvill (1967) and Weller (1967). The
following sections indicate major trends in country planning, working from
the microstructures of the countryside, such as fields and farms, through
settlement studies, to integrated planning in the countryside.

FIELD AND FARM CONSOLIDATION

In many European countries the task of field consolidation is an essential
first step if agricultural viability is to be achieved in the future. In open-
field areas the legacy of numerous, scattered tiny strips of which functioning farm units are composed has to be rationalized in order to facilitate
the mechanization of farming. The time wasted and the expense involved
in moving stock and implements between scattered strips is enormous
(Chisholm, 1962, p. 49). Most countries of western Europe now have schemes
in progress for various forms of consolidation (Lambert, 1963).

For example, in France the process of *remembrement* aims at regrouping
scattered ownership plots into a smaller number of blocks. Its *travaux connexes*
include the provision of new access routes and the removal of surplus field
boundaries. *Remembrement* is not concerned with farm enlargement. The

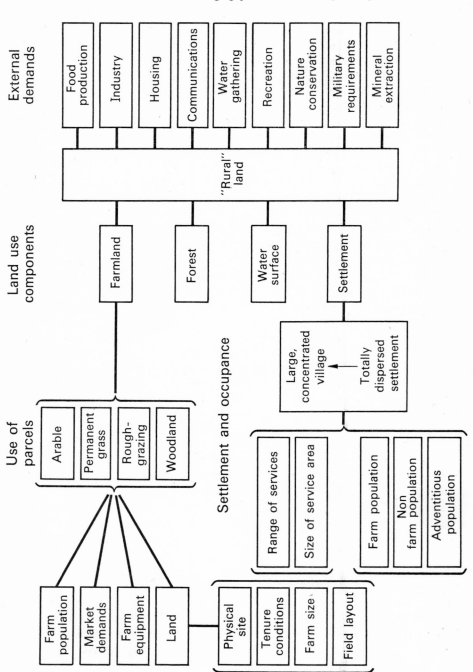

Fig. 20. Some components of country planning studies.

request for this improvement has to originate from local landowners who leave the practical task of consolidation to agricultural planners operating with financial aid from the State (Baker, 1961). The initiative for consolidation has been taken already in the richer agricultural regions, but now *remembrement* is spreading as landowners in poorer farming areas seek to emulate their neighbours' successes. Brutal changes, such as motorway construction which threatens to cut a farm into two isolated halves, have provided a more immediate incentive for reorganization. However, the French style of property consolidation is only a first step in the process of rural rationalization.

In other countries land consolidation has been taken a stage further. In the Netherlands, reorganization includes not only property consolidation but also farm enlargement and the construction of new farm buildings in keeping with modern needs (Lambert, 1961). For example, the destruction of farmhouses in the Maas Valley and the flooding of Walcheren Island during the Second World War provided the impetus for "internal reorganization" in these two areas with rectangular ring-fenced farms being laid out, together with new farmhouses and access roads (De Soet, 1962). But farm consolidation is easier in devastated or newly reclaimed areas than in those where traditional agriculture continues to function. The recently established *Sociétés d'Aménagement Foncier et d'Établissement Rural* (S.A.F.E.R.) in many regions of France acquire farmland as it comes on to the market, store it in a "land bank" and then use it to enlarge existing farms (Clout, 1968a). However, the resulting new farms are, on average, only one-third larger than their predecessors. Only a humble start has been made to modify an agricultural structure which becomes archaic as economic and technological conditions change.

Even in totally planned rural environments, farm structures have had to be enlarged in response to the demands of mechanization and the decreasing agricultural labour force. This may be exemplified by contrasting two polder areas in the Netherlands. The Noordoost Polder, drained in 1942, was divided according to differences in soil conditions into farm units ranging from 12 to 48 ha. By 1957 it was clear that such a size range was no longer suitable and hence the minimum farm size for the newly reclaimed Oostelijk Flevoland Polder was set between 15 and 20 ha (Constandse, 1961).

SETTLEMENT PLANNING AND SERVICE PROVISION

Newly reclaimed areas, such as the polders, and "old" rural regions where thorough-going rural reforms have been implemented have provided testing grounds for various types of settlement scheme. In several regions of Italy, for instance, there has been a marked dispersal of new farmsteads away from the traditional, densely-packed "agro-towns". This policy was deliberately chosen to wed formerly landless peasants to their newly-acquired farms and thus create a "rural democracy" (Desplanques, 1957). The disadvantages of settlement dispersal, such as increased costs of service provision and the loneliness experienced on the isolated farm (Chisholm, 1962, p. 136) have been reduced by grouping farmhouses in blocks of four at the focal point of four farm areas. Such small groupings involve only one building-site for constructing four farmhouses. A single well can serve four families and there is the possibility of help between neighbours in time of need. New, higher-order settlements ("colonization" and "service" centres) have been construc-ted to provide basic services and break the hold of the congested "agro-towns" over the surrounding countryside (Desplanques, 1957).

The planning of the Dutch polderlands indicates the rapidly changing conceptions of the "ideal" size and distribution pattern of rural settlements. In addition to 1600 farms set out in the midst of their fields, ten villages were established to house farmworkers and other personnel in the Noordoost Polder. The maximum distance between farm and village was 5 km. The planned "village region" covered some 4000 ha, by comparison with the average mainland "village region" of 1200 ha, and was designed to contain 3000 inhabitants, of whom about half would live in the village. Larger service centres were planned for populations of 10,000. But the scale of this operation proved to be incorrect and the scheme has not been successful. Many "village regions" have not attracted 3000 inhabitants. Farm workers are willing to travel greater distances to work than was previously believed. As a result of failing to attain these population targets, the village schools are too small to be efficient, church and voluntary organizations are not flourishing, and shopkeepers, faced with relatively few consumers, stock only a small range of goods (Constandse, 1961). A process of "ruralization" is in evidence as services which have been installed disappear and the remaining population becomes increasingly farm-based (Pinchemel, 1957).

The Dutch situation provides an accelerated illustration of what has happened more slowly in most rural areas. The village as a unit of settlement, service provision, and local administration was well suited to past conditions of greater rural population densities, in which face-to-face contacts were all important in the absence of generalized mobility (Mendras, 1965). But total village populations have decreased and many of the folk who remain can travel easily by private or public transport. Hence a case can now be made for running down the few costly and inefficient services which remain in most villages and concentrating them in "key villages" which, with the aid of adequate public transport, would serve much larger rural areas than the existing "village regions". As a result of the Noordoost Polder experience, the larger Oostelijk Flevoland settlement scheme includes only four villages aiming at populations of 3000 in "village regions" of 9000 ha which, in turn, are made up of larger farms (Constandse, 1961). In keeping with increased personal mobility, the maximum distance between farm and village has been increased to 10 km.

These examples have been laid out on a *tabula rasa* and are thus exceptions to the general rule. In long-settled countries, the "key village" policy would be painful to implement since existing small centres would be condemned to death. In the French context, at least, many planners fear that such a policy could lead to an undesirable volume of rural depopulation in peripheral areas. Thus the country planner is faced constantly with the discrepancy between systems of rational economic costing and social costing in the provision of rural services, for broader policies may define the desirability of maintaining population in the countryside, even though it will have to be provided with services which will never be viable.

One side to this planning dilemma is considered in an interesting theoretical plan for contraction of settlement in a remote area (Green, 1966). After an analysis of existing service provisions in rural Norfolk, a population of 5000 was considered to be the minimum size of settlement needed in the future to support a reasonable range of facilities, such as essential shops (including a chemist), a two-doctor medical practice, and a two-stream primary school. This could only be achieved by reducing development to an absolute minimum in nine parishes out of ten. The rapidity of scale changes is exemplified by the fact that 25 years previously figures between 1500 and 2500 were being quoted as an acceptable minimum size for a "healthy village" in lowland England (Orwin, 1944). In the current French context a size range of between 3000 and 10,000 inhabitants has been quoted for rural key villages (Guichard, 1965). Settlements which are easily accessible from surrounding communes and are the most culturally and economically successful may be designated as "village-centres" to be re-equipped to contain a junior school together with basic services, sports and cultural facilities (Di Borgo, 1966). Such a policy may be rational for country areas containing smaller and generally more mobile populations than in the past but its implementation will not be easy. The 38,000 communes in France " . . . have been inherited from the age of the bullock cart", and, although there is a need for regrouping into 2000 or so units, it is difficult to imagine any proud village community committing suicide (Moulin, 1968, p. 11). Sometimes it has been more a matter of murder, with forced grouping and resettlement schemes being imposed as a matter of military expediency. The resettlement in new villages of Algerian farmers at often unsuitable sites during the "war of liberation" exemplifies this form of hasty and ill-conceived rural "planning" (De Planhol, 1961a and 1961b; cf. Bourdieu and Sayad, 1964).

INTEGRATED PLANNING IN THE COUNTRYSIDE

The actions of the Highlands and Islands Development Board in Scotland, and the various French country planning companies, embody an integrated approach in which regional economies, land-use patterns, and demographic situations are analysed from fieldwork and statistical sources (Highlands and Islands Development Board, 1967; Clout, 1968b). As reliable numerical information is acquired in greater volume and depth, new data-processing equipment will permit more rapid and yet more exhaustive analyses of rural problems. In areas of contracting agriculture, such as the French Massif Central, two particularly important forms of detailed analysis are undertaken. First, the demographic trends of farming families within designated study areas are investigated to estimate the number of future farmers who are likely to remain on the land in this area of rapidly increasing rural depopulation (Fig. 21). After projections have been made, a plan is drawn up indicating lines of action which might be taken for land consolidation and farming rationalization. The study areas are then divided into "optimum land use zones" in which, for example, poor soils, steep slopes which cannot be used for mechanized farming, and areas remote from settlements, may be designated as "potential forest areas", where special efforts are made to encourage landowners to undertake afforestation. Areas of better land are designated as "potential agricultural zones" in which land improvements, such as *remembrement*, farm enlargement, and the construction of new buildings would be concentrated. "Intermediate zones" are outlined to accommodate deviations from the anticipated conditions. Many of the French country plans which result from such studies are multipurpose in character and include not only agriculture and forestry but also tourism. For instance, reservoirs and lakes behind hydro-electric barrages may be designated as focal points of new tourist development areas. A similar proposal has been made for parts of rural Wales (Hilling, 1968)

Industry may also fall within the realm of activity of the French country-planning companies, as in Languedoc, but the jobs created often fail to meet the crucial need for new forms of *male* employment. Research in mid-Wales has provided a useful statement on industrial movements to rural areas (H.M.S.O., 1964). It was found that industrialists generally moved to existing premises, and that few would start *ab initio*; their units were small and generally employed women; growth prospects for such concerns were small. Indeed, French rural industrial experience has shown that local labour forces were often unsuitable and workers have had to be brought in from outside while the local population was trained. Transport costs for raw materials and finished goods were higher in remote rural areas and often implanted firms were "branch" concerns which would be closed first in times of economic

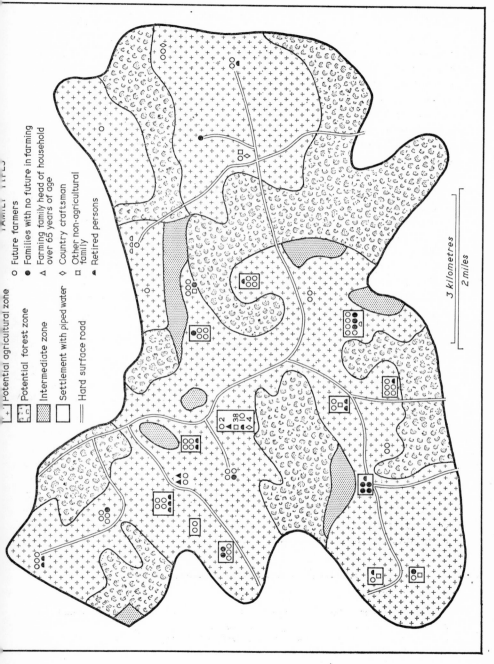

Potential agricultural zone ○ Future farmers
Potential forest zone ● Families with no future in farming
Intermediate zone △ Farming family head of household over 65 years of age
Settlement with piped water ◇ Country craftsman
Hard surface road □ Other non-agricultural family
▲ Retired persons

3 kilometres
2 miles

FIG. 21. An example of a French rural planning map.

hardship. In Wales "the most important inhibiting factor from the industrialist's point of view is the shortage of labour" linked to the dispersal of population, for "it is probably true to say that the settlement pattern, combined with a low level of population, is the basic cause of many of the problems of mid-Wales" (H.M.S.O., 1964, p. 5). This is true of most rural areas.

The mid-Wales report does not stress the possibility of expanding tourism which, by contrast, is emphasized in French rural planning studies as an important means of diversifying local economies and bringing back, if only seasonally, some of the capital which has drained townwards with rural depopulation (Guichard, 1965). The presence of itinerant tourists and residents in country cottages can be important in generating seasonal employment and extra incomes in rural areas, but it can also place severe strains on local finances and on services such as water supplies (Cribier, 1966). Some agriculturalists have become veritable *vendeurs de vacances* rather than full-time farmers (Rambaud, 1967) and as a result the life of the village and its social psychology will be altered by the influence of the outside world—by the new orientation of the economic life of its rural community (Saville, 1966, p. 32). New economic priorities have been established, new work and leisure patterns are developing and new social structures are appearing (George, 1964 and 1965; Anderson, 1963).

CONCLUSION

Agriculture is supporting an ever-decreasing proportion of total national populations, but this does not mean that planning studies in rural areas do not merit our close attention. These reports emphasize the increasing discordance between rapidly changing social, economic, and land-use conditions and the historically inherited environments within which they are uncomfortably set. The detailed physical and cultural investigations which are now being undertaken as part of integrated rural planning open a dynamic research field covering such major themes as locational analysis and the diffusion of innovations. The need to plan the countryside presents a challenge to the geographer to apply his knowledge to rural problems so that new demands may be met with a minimum of destructive chaos and for the maximum benefit of the greatest number.

REFERENCES AND FURTHER READING

INTRODUCTION

Valuable surveys of country planning problems in the British context are contained in:
ARVILL, R. (1967) *Man and Environment,* Penguin, Harmondsworth.
ORWIN, C. S. (Ed.) (1944) *Country Planning,* Oxford U.P., London.

WELLER, J. (1967) *Modern Agriculture and Rural Planning*, Architectural P., London.

Aspects of rural sociology are considered in:

FRANKENBERG, R. (1966) *Communities in Britain*, Penguin, Harmondsworth.
PAHL, R. (1965) Class and community in English commuter villages, *Sociologia Ruralis*, **5**, 5–23.
PAHL, R. (1966) The rural–urban continuum, *Sociologia Ruralis*, **6**, 299–329.

Additional information on these topics is presented in:

BEST, R. H. and COPPOCK, J. T. (1962) *The Changing Use of Land in Britain*, Faber, London.
HIGGS, J. (Ed.) (1966) *People in the Countryside—studies in rural social development*, National Council of Social Service, London.

FIELD AND FARM CONSOLIDATION

Considerations of consolidation policies are presented in:

BAKER, A. R. H. (1961) Le remembrement rural en France, *Geography*, **46**, 60–2.
CHISHOLM, M. (1962) *Rural Settlement and Land Use*, Hutchinson, London.
CLOUT, H. D. (1968a) Planned and unplanned changes in French farm structures, *Geography*, **53**, 311–15.
CONSTANDSE, A. K. (1961) Planning in agricultural regions, *Sociologia Ruralis*, **2**, 79–104.
LAMBERT, A. (1961) Farm consolidation and improvement in the Netherlands, an example from the Land Van Maas en Waal, *Economic Geography*, **37**, 115–23.
LAMBERT, A. (1963) Farm consolidation in Western Europe, *Geography*, **48**, 31–48. (Contains a full bibliography.)
DE SOET, D. (1962) *Rural Development in the Netherlands*, Ministry of Agriculture and Fisheries, The Hague.

Additional information and specific examples may be found in:

THOMPSON, I. B. (1961) Le remembrement rural; a case study from Lorraine, *Geography*, **46**, 240–2.
WHITE, J. T. (1966) Kerastren Bras, a private remembrement rural, *Geography*, **51**, 246–8.

SETTLEMENT PLANNING AND SERVICE PROVISION

DI BORGO, C. P. (1966) A French experiment in rural development, in HIGGS, J. (Ed.) *People in the Countryside—studies in rural social development*, National Council of Social Service, London, pp. 109–32.
BOURDIEU, P. and SAYAD, A. (1964) Paysans déracinés, *Etudes Rurales*, **12**, 56–94.
DESPLANQUES, H. (1957) La réforme agraire italienne, *Annales de Géographie*, **66**, 310–27.
GUICHARD, O. (1965) *Aménager la France*, Laffont-Gouthier, Paris.
GREEN, R. T. (1966) The remote countryside—a plan for contraction, *Planning Outlook*, **1**, 17–37.
MENDRAS, H. (1965) *Sociologie de la Campagne Française*, P.U.F., Paris.
MOULIN, J. (1968) *Les Citoyens au Pouvoir—12 régions, 2,000 communes*, Le Seuil, Paris.
PINCHEMEL, P. (1957) *Structures sociales et dépopulation rurale dans les campagnes picardes de 1836 à 1936*, Colin, Paris.
DE PLANHOL, X. (1961a) Les nouveaux villages d'Algérie, *Geografiska Annaler*, **43**, 243–51.
DE PLANHOL, X. (1961b) *Nouveaux Villages Algérois*, P.U.F., Paris.

Additional information may be found in:

BRACEY, H. (1952) *Social Provision in Rural Wiltshire*, Methuen, London.
HOUSE, J. W. (1965) *Rural North-east England, 1951–1961*, Department of Geography, Newcastle.

INTEGRATED PLANNING IN THE COUNTRYSIDE

ANDERSON, N. (1963) Aspects of the rural and urban, *Sociologia Ruralis*, **3**, 8–22.
CLOUT, H. D. (1968b) France renovates her rural areas, *Town and Country Planning*, **36**, 312–61.
CRIBIER, F. (1966) 300,000 résidences secondaires, *Urbanisme*, **96-97**, 97–101.
H.M.S.O. (1964) *Depopulation in Mid-Wales*, 1964, H.M.S.O., London.
GEORGE, P. (1964) Anciennes et nouvelles classes sociales dans les campagnes françaises, *Cahiers Intérnationaux de Sociologie*, **37**, 13–21.
GEORGE, P. (1965) Quelques types régionaux de composition sociale dans les campagnes françaises, *Cahiers Intérnationaux de la Sociologie*, **38**, 49–56.
Highlands and Islands Development Board (1967) *First Report*, H.I.D.B., Aberdeen.
HILLING, J. B. (1968) Mid-Wales; a plan for the region, *Journal of the Town Planning Institute*, **54**, 70–4.
RAMBAUD, P. (1967) Tourisme et urbanisation des campagnes, *Sociologia Ruralis*, **7**, 311–34.
SAVILLE, J. (1966) Urbanisation and the countryside, in HIGGS, J. (Ed.) *People in the Country-side—studies in rural social development*, National Council of Social Service, London, pp. 13–34.

Further examples of integrated planning schemes and their components may be found in:

CLOUT, H. D. (1969) Country planning in Gascony, *Scottish Geographical Magazine*, **85**, 9–16.
GRAVES, N. (1965) Une Californie Française—the Languedoc and Lower Rhône irrigation project, *Geography*, **50**, 71–3.
THOMPSON, I. B. (1966) Some problems of regional planning in predominantly rural environments; the French experience in Corsica, *Scottish Geographical Magazine*, **82**, 119–29.

CHAPTER 22

GEOGRAPHY AND
LOCAL GOVERNMENT REFORM

N. H. PERRY

APPLIED geography, as defined in the Second Anglo-Polish Seminar (Leszc-zycki, 1964; Wise, 1964), is concerned with problems of the real world. It does not have any distinctive methodology but uses the general body of geographical techniques to solve practical problems. Within this definition of applied geography, local government is a field which has received attention from geographers for well over half a century and has recently become highly topical. The late 1960's are witnessing a major restructuring of British local government. The Royal Commission on Local Government in England has reported, a Commission is sitting for Scotland, and a White Paper has been published on Wales.

THE FUNCTIONS OF LOCAL GOVERNMENT

Any discussion of local government reform must be preceded by an appreciation of the functions of local government. Why does it exist? What sort of government does it or should it provide, and how "local" is it or should it be? The functions exercised by local authorities in Britain include housing, town-planning, education, police, fire service, public transport, environmental health services, water-supply, vehicle licensing, highways, parks, weights and measures, abattoirs, and an ever-widening range of social services. Before 1947, gas and electricity could often have been included in the list. Despite the apparently growing power of central government, the range of activities performed at a local authority level has consistently expanded. The scale at which some or all of these services have been administered has ranged from the parish through rural and urban districts, counties, and county boroughs to the newly proposed unitary areas.

As well as providing an areal base for the provision of manifold services, local government units have been regarded as the primary cells of the political system, as the repositories of proud historical traditions which are

233

especially well-pronounced at a county and city level, and as the manifestations of deeply felt local patriotism. Any scheme of local government reform must therefore take into account a whole suite of interrelated criteria.

THE CHOICE OF CRITERIA
FOR LOCAL GOVERNMENT REFORM

Local government reform cannot be considered in isolation from the balance of national political power. In theoretical terms this has found expression in discussions of the "optimum" number of strata for the organization of government. Some consider that the optimum number of levels is three, with the national government providing the top stratum, local administration the bottom, and the province or county constituting an intermediary layer (Ostrom *et al.*, 1961; Ylvisaker, 1964). The number of administrative levels chosen is basic to local government reform and profoundly influences later decisions.

Having decided the number of levels, a choice must then be made as to how far functional efficiency should be sacrificed in the interests of political objectives. Given the two general propositions that (1) maximum functional efficiency is desirable, and (2) due regard must be paid to political imperatives, the historical context, and local emotional ties, likely approaches to a solution might involve the following considerations.

First, one might seek a solution yielding the maximum efficiency as measured by financial yardsticks, and tempered by political and emotional criteria to a minimal extent. Or, on the other hand, it might be considered that the political acceptability of a solution is the paramount concern and that, within this limitation, as much efficiency as possible should be sought. It is vital that a conscious choice should be made between functional efficiency and political acceptability, if they prove not to be completely compatible. If a compromise has to be made between them, its nature must be clearly stated. If there has to be a sacrifice of functional efficiency in the interests of less tangible criteria, then some estimate of the financial costs should be attempted. If the level of citizen participation in government is to be restricted in the interests of "efficient" administration, where will restrictions be imposed and how strong should they be allowed to become? It is important that political leaders, in whose hands the decisions lie, should become used to this sort of thought process, so essential is it to rational decision-making.

THE RÔLE OF RESEARCH

If they are to make rational decisions about local government organization, politicians must be provided with the factual and analytic data on which to base their thinking. With this aim in mind, applied geographers and their colleagues from other disciplines are extending the amount of information on the behaviour and operation of local government systems. These studies are taking on an increasingly multi-disciplinary flavour.

Political scientists and sociologists, for example, are testing the operation of political systems at various levels of government and investigating the extent to which citizens participate or wish to participate in the governmental process. Working in Canada, Kaplan (1967) has studied the operation of the Metropolitan Toronto government during its first few years, with a view to assessing its performance in terms of the "functionalist" sociology pioneered by Talcott Parsons (1951). This approach to sociology examines the behaviour of social "systems". Each system has a series of "norms" (standards of behaviour) and a number of "rôles" (recognizable patterns of behaviour). The degree to which a system runs smoothly according to its norms and the rôles being played is termed its "integration". In his study of the integration of the Toronto system, Kaplan raised some interesting questions. It is true to say that larger political units decrease effective democracy? Is it perhaps truer that like-minded people have a better chance of forming bigger groups and getting things done in larger political units? Is it, perhaps, also true that the effective local unit of people's participation in affairs is so restricted anyway (about the size of a street or block) that it is too small for any administrative purposes and, above that level, people are often indifferent to size? The answers to these possibly tendentious questions are not known, but any scheme of local government reform which sought to be politically sensitive would have to take such factors into account.

Economists and operations researchers are more concerned with efficiency in operating local authority services. Of particular importance is the scale at which local government functions can be efficiently operated. These are often linked with population totals. Certain critical threshold costs exist for the supply of water, gas, and electricity and for the operation of sewage disposal and other public utilities. The Royal Commission on Local Government in England was aware of this problem and commissioned research on economies of scale in local government services (Institute of Social and Economic Research, 1968; Hutton and Gupta, 1968) and on the social aspects of local authority services (Government Social Survey, 1968). Koslowski (1968), in studies connected with the Grangemouth/Falkirk Sub-Regional Plan, used Polish experience in threshold theory to demonstrate the usefulness of this theory in determining the size of administrative areas. A further factor arises from the civil service practice of laying down certain salary grades for local government units in specific population ranges. Within this framework, the best possible deployment of the available pool of local government officers must be attempted.

The geographer is usually concerned with the spatial aspects of functional efficiency in local government. Most workers have taken it for granted that local government units will continue to be contiguous, strictly delimited, general-purpose administration areas. Within this context a search has been made for optimum areal sizes of local government units on functional grounds. Work has also been done on the shapes of local government units and investigations made in order to determine and/or verify the existence of a network or hierarchy of activities which could be used as the base for local government structure (Bunge, 1962; Boyce and Clark, 1964). Identifications of such hierarchies has been sought on the basis of shopping facilities, public transport linkages, and journeys to work. Studies of the current operation of local government on a spatial basis can suggest ideas on the level of functional efficiency attained or attainable.

GEOGRAPHICAL RESEARCH ON LOCAL GOVERNMENT

Geographical work on the topics mentioned above and in similar fields was stimulated in Britain during the Second World War, when various schemes of regional division were required for wartime administration and when the ideas of Walter Christaller on the hierarchical arrangement of urban activities had filtered through to research workers in this country and had been partially digested (Christaller, 1933; Dickinson, 1932). In particular, Smailes (1944 and 1947) used indices of shopping, cultural, educational, and administrative provision to indicate a hierarchy of urban places in England and Wales. This work had two main aspects. On the one hand, it had to be decided what sort of services characterized different levels of urbanism and what type of threshold values between categories should be chosen. On the other hand, the problems of demarcating boundaries to the areas of influence of individual urban centres had to be faced.

Most subsequent work was also based on the twin problems of urban centrality and boundary definition. A division can be made between work which used "traditional" cartographic techniques and more recent research employing some of the new armoury of statistical and mathematical techniques. In the former category can be included Green's maps of accessibility by bus services. Although relevant at the time when the research was done, the rise in car-ownership and personal mobility has robbed the maps of some of their value (Green, 1950 and 1966). Carruthers, of the Ministry of Housing, has published a series of papers concerned with service centres in England and Wales, and shopping centres in Greater London and in England and Wales (Carruthers, 1957, 1962, and 1967).

The lead in using newer analytical techniques has often come from North America. For example, Zobler introduced the use of Chi-square techniques for the statistical testing of regional boundaries and Curry, employing ideas originally put forward by August Lösch, used statistical methods and operations research techniques in the study of service centres within towns (Zobler, 1957; Curry, 1962; Lösch, 1944). This lead has been taken up in Britain, though not always by geographers. Mills used linear programming methods to define electoral wards in Bristol under certain assumptions of population size and ward area (Mills, 1967). In classifying British towns by their social and economic characteristics, Moser and Scott used the technique known as "principal components analysis" to assess the relative importance of various factors; Gittus applied similar techniques for the subdivision of urban areas (Moser and Scott, 1961; Gittus, 1964). Statistical techniques and sociological theory have been used by Robson in his study of social change in Sunderland (Robson, 1966).

Research by geographers into problems of local government has by no means been confined to the English-speaking world. In recent years European geographers, faced with the new realities of the post-war years and the rapid changes in national, international and supranational relationships, have responded to the challenge. The work of German geographers was encouraged by the provisions which the West German Constitution makes for a redrawing of internal boundaries. Early work by Hartke (1948) on administrative problems in the Rhine–Main area of southern Germany considered the rôle which newspaper circulation, journey to work data, and the provision of public services could play as criteria in administrative reform. The Luther Commission, set up by the Federal Government in the early 1950's to consider the boundaries of the West German *Länder*, commissioned a major piece of work from Schöller (1966). He tested the allegiance of people living in the Westerwald to the various *Länder* (North Rhine–Westphalia, Hessen, Rheinland–Pfalz) between which the area is split and made recommendations on the basis of his investigations. Tietze (1965) also commented on the general problems of local government boundaries in Germany.

French interest in physical planning and related problems of local government has become lively in recent years. The creation of various systems of special-purpose regions for administrative purposes has encouraged research workers to seek a more rational basis for territorial division (Lanversin, 1965; Club Jean Moulin, 1968).

PROPOSALS BY GEOGRAPHERS
FOR LOCAL GOVERNMENT REFORM

Given the existence of a body of geographical work on some of the basic problems relevant to local government organization, it is hardly surprising that some geographers have felt themselves able to offer practical solutions. Examination of some of these reveals the extent to which political and functional criteria have, explicitly or intuitively, been taken into account.

Prescott (1965), reviewing work by Fawcett (1919), Gilbert (1939 and 1948), and Taylor (1942), noticed that the authors agreed on four proposals:

(1) All government areas should be composed of aggregates of the smallest basic units, grouped in such a way that their boundaries will be multi-functional.

(2) As far as possible, each main administrative area should include related urban and rural areas, and residential and industrial districts.

(3) Individual conurbations should constitute single administrative areas, capable of co-ordinating development within and between them.

(4) Boundaries should be drawn to cater for local sentiment and regional patriotism.

Fawcett was alone in recommending that boundaries should be drawn through lightly populated or uninhabited areas and should follow watersheds, thus avoiding unnecessary divisions of water, road, and sewage services, which often follow valleys.

These proposals combined a rich mixture of functional and political requirements with certain implicit contradictions. For example, the concept of the "smallest basic unit" was never clearly defined, but had links with the "neighbourhood" idea applied in the planning schemes of Sir Patrick Abercrombie. Again, the call for unified conurbation authorities could often clash with local sentiment. But, in general, the proposals certainly showed an awareness of the need to maintain political credibility while adapting administration to the rapidly changing structure of urban regions which was clearly emerging, even in pre-war days.

OTHER PROPOSALS FOR LOCAL GOVERNMENT REFORM

Proposals by non-geographers for schemes of local government reform in the last 20 years have been numerous. These proposals have come from private and official or semi-official sources.

Private proposals for reform have often been sponsored by political parties or organizations. For example, a Fabian booklet by Self (1949) highlighted the situation as seen by many people at a time when the apparatus of wartime government and post-war social administration appeared to pose dangers for local government. Self's main aim was to ensure adequate government for conurbations, while maintaining and re-invigorating local democracy. His great fear was of

> ... the new brand of *ad hoc* regionalism, typified by the regional boards for hospitals, gas and electricity. Instead of being a healthy upward growth from below, these new limbs stretch downward from the swollen trunk of national administration.

Self's comments have acquired fresh relevance with the formation of new planning regions in the early 1960's, which are tied to central government. Other private schemes for reform have included that of the Liberal Party, proposing the division of England into twelve provinces, broadly on the same lines as those proposed by Fawcett in 1919, and the scheme put forward by D. Senior at the Anglo-American Planning Seminar in 1965 (Senior, 1966).

Official and semi-official proposals for reform also have a long history, which can be traced back to the creation of the Poor Law Unions in 1834 (Freeman, 1968). Before the Royal Commission on Local Government in England was set up in 1966, the main agent of local government reform was the Local Government Commission established in 1958. This commission toured the country, making proposals for reform within very limiting constraints laid down by the Government. Its work made full use of conventional geographical techniques: for example, the report on Tyneside included maps of the organization of selected local government services and of the journey to work pattern (Local Government Commission, 1963). In general, the recommendations of the commission supported the creation of larger local government units.

CONTEMPORARY OPINION

The work of the Royal Commission prompted a major rethinking of the form of local government. Many interested organizations submitted evidence and it is clear from this evidence that the consensus of contemporary opinion favours larger units (Gowan and Gibson, 1968). Analysis of 118 memoranda submitted to the Royal Commission reveals five types of proposed solution (Fig. 22):

(1) *A full regional system, based on the economic planning regions.* In this system, the economic planning regions would have powers and duties similar to those already existing for the Greater London Council. The lower-tier authority would resemble the London boroughs in urban areas and enlarged rural districts elsewhere, with the emphasis on lower-tier flexibility;

(2) *A two-tier system, with a top tier of city regions.* Newly constituted city regional authorities would exercise most or all of the major local government functions, with only a weak structure of lower-tier authorities;

(3) *A two-tier system, with present counties as the top tier.* Essentially a modification of the present local government apparatus, this solution would amalgamate the present counties and county districts into bigger groupings;

(4) *Single-tier city regions.* In this system, government in England would effectively be reduced to two levels, the central government in London

Proposals for local government re-organization

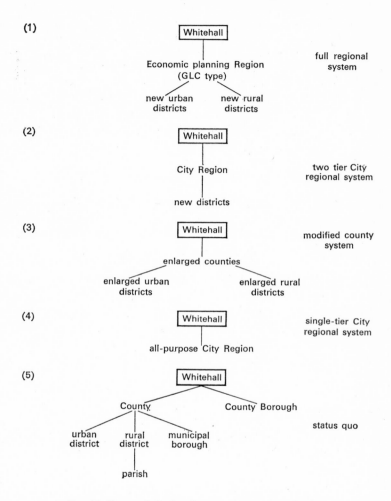

FIG. 22. Proposals for local government reorganization.

and a number of large and powerful city regional authorities covering the whole country. These authorities would exercise all the local government functions, which are at present split between different levels of county administration;

(5) *Preservation of the status quo.* Only a small number of witnesses before the Royal Commission, of which the most prominent was the City of

Birmingham, saw no faults in the present system which necessitated reform.

The most powerful body of opinion, that of the Government departments, supported the formation of some thirty to forty major city regional authorities. A point to notice is that all the proposals were conditioned by the relationship between potentially large local government areas and the state. Although functional criteria received much attention in evidence given to the commission, far less thought was given to the implications of relationships between the authorities who would govern city regions and the people who live in them.

CONCLUSION

Although most serious proposals for local government reform have at least observed the existence of tension between political and functional forces, few have offered solutions which are based on much more than mere intuition. Future geographical work on local government will have to abandon this intuitive approach and align itself more closely to the more analytical techniques pioneered by sociologists, economists, and operations researchers. There need be little danger of geography losing its individuality in this process. Provided that the opportunities are grasped, geographers, with their deeply rooted involvement in spatial relationships, are capable of providing genuine insights into the processes of government. This observation is as true of geography teaching as it is of research. Demonstrations of geography's methods and intellectual approach at work in "real" situations has, at the same time, intrinsic interest, and can heighten awareness in students. The connection with local government forges a valid link with the community in which pupils live and increases the contribution which geography can make towards satisfying the increasing interest in socal problems shown by students.

REFERENCES AND FURTHER READING

APPLIED GEOGRAPHY AND LOCAL GOVERNMENT

Recent general books on geography and local government:

FREEMAN, T. W. (1968) *Geography and Regional Administration*, Hutchinson, London.
MACKINTOSH, J. P. (1968) *The Devolution of Power*, Penguin, Harmondsworth.

Definitions of applied geography are given in:

LESZCZYCKI, S. (1964) Applied geography or practical applications of geographical research, *Geographia Polonica*, **2**, 11–22.
WISE, M. J. (1964) The scope and aims of applied geography in Great Britain, *Geographia Polonia*, **2**, 23–36.

THE CHOICE OF CRITERIA FOR LOCAL GOVERNMENT REFORM

Two articles to illustrate the theoretical principles underlying the organization of government were reprinted in:

FRIEDMANN, J. and ALONSO, W. (1964) *Regional Development and Planning: a reader*, M.I.T. P., Cambridge, Mass.

they are:

OSTROM, V. *et al.* (1961) The organization of government in metropolitan areas—a theoretical inquiry, pp. 542–53.

YLVISAKER, P. (1964) Some criteria for a "proper" areal division of governmental powers, pp. 519–41.

THE ROLE OF RESEARCH

Examples of the work of sociologists on local government matters include:

Government Social Survey (1968) *Local Authority services and the characteristics of administrative areas*, H.M.S.O., London.

KAPLAN, H. (1967) *Urban Political Systems*, Columbia U.P., New York.

PARSONS, T. (1951) *The Social System*, The Free P., Glencoe.

Economic and operations research work relevant to local government:

HUTTON, J. P. and GUPTA, S. P. (1968) The size of local authorities, *New Society*, 4th July.

Institute of Economic and Social Research, University of York (1968) *Economies of Scale in Local Government Services*, U. of York P., York.

KOSLOWSKI, J. (1968) Threshold theory and the sub-regional plan, *Town Planning Review*, **39**, 99–116.

MILLS, G. (1967) The determination of local government electoral boundaries, *Operational Research Quarterly*, **18**, 243–55.

MOSER, C. and SCOTT, W. (1961) *British Towns*, Oliver & Boyd, Edinburgh.

Geographical work relevant to local government can be broadly divided into (1) the earlier work, using "traditional" techniques of investigation, and (2) newer, more quantitatively-oriented research.

(1)

CARRUTHERS, W. I. (1957) A classification of service centres in England and Wales, *Geographical Journal*, **123**, 371–85.

CARRUTHERS, W. I. (1962) Service Centres in Greater London, *Town Planning Review*, **33**, 5–31.

CARRUTHERS, W. I. (1967) Major shopping centres in England and Wales 1961, *Regional Studies*, **1**, 65–81.

CHRISTALLER, W. (1933) *The Central Places of Southern Germany*, English translation, 1966, Prentice-Hall, Englewood Cliffs.

DICKINSON, R. E. (1932) The distribution and functions of the smaller urban settlements of East Anglia, *Geography*, **7**, 19–31.

GREEN, F. H. W. (1950) Urban hinterlands in England and Wales: an analysis of bus services, *Geographical Journal*, **116**, 64–88.

GREEN, F. H. W. (1966) Urban hinterlands: fifteen years on, *Geographical Journal*, **132**, 263–86.

SMAILES, A. E. (1944) The urban hierarchy in England and Wales, *Geography*, **29**, 41–51.

SMAILES, A. E. (1947) The analysis and delimitation of urban fields, *Geography*, **32**, 151–61.

(2)

BOYCE, R. B. and CLARK, W. A. V. (1964) The concept of shape in geography, *Geographical Review*, **54**, 561–72.

BUNGE, W. (1962) Theoretical geography, *Lund Studies in Geography, series C,* **1,** revised edition, 1966.

CURRY, L. (1962) The geography of service centres within towns, in NORBORG, K. (Ed.) *Proceedings of the I.G.U. Symposium in Urban Geography, Lund 1960,* Gleerup, Lund, pp. 31–53.

GITTUS, E. (1964) The structure of urban areas: a new approach, *Town Planning Review,* **35,** 5–20.

LÖSCH, A. (1944) *The Economics of Location,* English translation, 1954, Yale U.P., New Haven.

ROBSON, B. T. (1966) An ecological analysis of the evolution of residential areas in Sunderland, *Urban Studies,* **3,** 120–42.

ZOBLER, L. (1957) Statistical testing of regional boundaries, *Annals of the Association of American Geographers,* **47,** 83–95.

German work mentioned in the text:

HARTKE, W. (1948) Gliederung und Grenzen im Kleinen: Rhein Main-Gebiet, *Erdkunde,* **2,** 174–9.

SCHÖLLER, P. (1966) Neugliederung, *Forschungen zur Deutschen Landeskunde,* **150.**

TIETZE, W. (1965) Zum Problem der Verwaltungsgrenzen in der deutschen Kulturlandschaft, *Festschrift Leopold G. Scheidl zum 60. Geburtstag, Teil I,* Berger & Söhne, Vienna, pp. 368–72.

French work mentioned in the text:

Club Jean Moulin (1968) *Les Citoyens au Pouvoir—12 Regions, 2000 Communes,* Editions Seuil, Paris.

LANVERSIN, J. DE (1965) *L'Aménagement du Territoire,* Libraries Techniques, Paris.

PROPOSALS BY GEOGRAPHERS FOR LOCAL GOVERNMENT REFORM

FAWCETT, C. B. (1919) *The Provinces of England,* Reprinted by Hutchinson, London, 1961.

GILBERT, E. W. (1939) Practical regionalism in England and Wales, *Geographical Journal,* **94,** 24–44.

GILBERT, E. W. (1948) The boundaries of local government areas, *Geographical Journal,* **111,** 172–206.

PRESCOTT, J. (1965) *The Geography of Frontiers and Boundaries,* Hutchinson, London.

TAYLOR, E. G. R. (1942) A discussion on the geographical aspects of regional planning, *Geographical Journal,* **99,** 61–80.

OTHER PROPOSALS FOR LOCAL GOVERNMENT REFORM

Private proposals for reform include:

SELF, P. (1949) *Regionalism,* The Fabian Society, London.

SENIOR, D. (Ed.) (1966) *The Regional City,* Longmans, London.

Official proposals for reform include:

GOWAN, I. and GIBSON, J. (1968) The Royal Commission on Local Government in England: a survey of some of the written evidence, *Public Administration,* **46,** 13–25.

Local Government Commission for England (1963) *Report and Proposals for the Tyneside Special Review Area, Report,* **5,** H.M.S.O., London.

Report of the Inspector appointed by the Minister of Housing and Local Government to Hear Objections to the Proposals of the Local Government Commission for England for the Tyneside Special Review Area, (1965), H.M.S.O., London.

Local Government in Wales (1968) *Cmnd. 3340,* H.M.S.O., London.

Royal Commission on Local Government in England, 1969, *Report, Cmnd.* 4040 H.M.S.O., London.

PART IV

AREA STUDIES

Youth is the time to go flashing from one end
of the world to the other both in mind and
body; to try the manners of different nations. . . .
R. L. STEVENSON, *Virginibus Puerisque*

CHAPTER 23

GEOGRAPHY AND AREA STUDIES

W. R. MEAD

ACADEMIC geography has gained many recruits because of the appeal of other lands. The exotic attraction is found at all levels. In the schoolroom it is the different character of other countries and of other peoples that captures youthful imagination. The immense circulation of the pictorial geographical magazines is inseparable from it; so is the broad membership of the geographical societies. There may be a minority of students who pursue geography objectively with an eye to a career, but for the most part the generous recruitment into departments of geography is subjective and inseparable from the curiosity aroused by distant places and their strange phenomena.

During the last two generations the study of foreign lands has been loosely classified as regional geography. In its original form, regional geography drew its facts from the entire systematic range of the subject, and attempted to resolve its findings in a final synthesis. Subsequently, the term "regional geography" has been widely misused. It has been misapplied in the titles as well as in the content of textbooks. Thus areas are described as regions regardless of whether they conform to the definition of a region; while a synthesis is rarely attempted from the miscellany of facts of which they are composed. Independently of these considerations, it is increasingly urged that the regional approach is by no means the most satisfying way to study foreign lands.

Although the method of traditional regional geography may be unsatisfactory, the geography of other lands remains a necessary part of general knowledge. One of the problems of contemporary geographers is that of developing, disciplining, and organizing this part of the subject so that it retains both appeal and respect. The phrase "area studies" has generally been adopted to cover it.

At the outset, it is important to make a distinction between the use of the phrase "area studies" in an organizational context and its employment to cover the detailed treatment of particular areas of the world. Organizationally, the phrase is associated with the younger universities which have initiated degrees under this title, and with the older universities which have created institutes of postgraduate studies focused on particular areas. The Schools of

European Studies at Sussex and East Anglia illustrate the former. The array of postgraduate institutes in the University of London (e.g. of Latin American Studies, of Commonwealth Studies, of United States Studies), and the Institutes of Russian Studies in Glasgow and South-east Asian Studies in Hull illustrate the latter. In each case they are institutes and schools of combined studies and, since they devote their attention to well-defined world areas, geography is logically central to them.

The concept of combined studies is a sensitive academic issue. The sensitivity is shared by the professional geographer no less than by the historian or the linguist. It is immediately argued that any form of combined studies represents a dilution of the component parts. Beyond this, academic geographers have tended to grow increasingly critical of those in their ranks who deal with the general to the exclusion of the particular. They have sounded "the retreat from cosmology": they have campaigned for increasing specialization. The growth of research and publication has led to pressures for the addition of a fourth year to degree courses, or to increasing concentration—even at the undergraduate level—on a limited area of the subject. If an honours geography student, who devotes his entire undergraduate life to the subject cannot cope with its essentials, critics of schools of area studies argue that it is impossible for an undergraduate to pursue a group of subjects simultaneously and emerge with anything remotely resembling a respectable qualification in any of them on graduation. The situation becomes even more debatable at the postgraduate level. Can professional geographers admit postgraduate students to read and be examined in a subject for a higher degree without having taken courses in the same discipline at the undergraduate level? At the very least, this causes a ruffling of professional feathers. Yet a compromise must be struck, because area studies without a geographical component are a contradiction in terms.

Area studies have a second and quite unambiguous connotation. The term covers geographical work pursued in a well-defined area of the world. The old term—regional geography—inexplicit in itself, also gives a false impression of the predominant methodologies currently employed. Many students concerned with area studies look first to the identification of significant problems and second to the search for solutions to these problems. In general, the approach to problems inclines towards practical and applied issues, although there will always be those who are drawn to past problems and the attempt to explain them. Formerly, it was the rounded education springing out of geography that was presented as the distinguishing feature of a training in the discipline. Contemporarily, it is the geographer with competence in a limited specialism who commands the stage: increasingly, the specialism tends to be systematic. Given expertise in a particular branch or branches of the subject, this can be appropriately adapted to particular places. In terms of numbers, systematic specialists with an interest in a particular area are taking precedence over area specialists with a systematic interest. Both have a place, though they are likely to employ their energies differently. The systematist with an area interest is likely to incline towards applied research; the area specialist with an interest in a systematic field is either likely to favour pure research or to function as an interpreter of other areas. Schools of area studies will probably grow at least as much around those who interpret, as around those who apply their systematic knowledge to specific problems.

Interpretation demands different equipment at different levels. In the schoolroom it must depend to a large extent on visual aids and on the teacher's ability to interpret the written experiences of other people. At the university, interpretation is more usually based on close personal experience of an area and its literature. There are those who have the ability to interpret faithfully and effectively from limited acquaintance. In general, however, interpretation calls for prolonged experience of an area. The problem is then of selection and arrangement from among the stream of personal impressions and from the endless flow of literature. In the exercise of interpretation the imagination must be restrained by the intellect, the private experience fitted into the context of public experience.

Whether the object of area studies is applied research or interpretation of a situation, the heart of the matter is the identification of a relationship between people and place. Most of the problems central to human geography concern the attachment of people to place and the causes that give rise to their activities in a particular place. At their simplest, these causes are often very personal. The associations that tie people to particular pieces of land or tracts of country have been curiously ignored by geographers, though they emerge from sensitive autobiography. Herbert Read (1963) touched upon them in his appreciation of the farm where he spent his boyhood in the Vale of York; so did Georges Duhamel (1931) describing his childhood associations with the Île de France. For D. H. Lawrence (1924), the association was explained in terms of the human response to a "different vital effluence, different vibration, different chemical exhalation, different polarity with a different star" of a particular piece of land. The American author Hamlin Garland (1963) identified a closer blood and soil relationship. Robert Ardrey (1967) has come nearer to the analysis of it than anyone in his debatable concept of the territorial imperative—the animal instinct that attaches man to a particular area of land. It is a concept which awaits scientific explanation, but which is critical for understanding nationality. Aldous Huxley (1963), sensitive to the need for harmony between men and the land where they live, looked at it from a different angle.

Men cannot live at ease except when they have mastered their surroundings and where their accumulated lives outnumber and outweigh the vegetative lives about them. Stripped of its dark woods, planted, terraced and tilled almost to the mountain tops, the Tuscan landscape is humanized and safe . . . one is glad to return to the civilized and submissive scene [Huxley, 1937, p. 39].

Such considerations are central to area studies—and are a fruitful field for combined studies.

They prompt several comments. First, students might be invited to join in an introspective exercise by trying to put into words the strength of their feeling for a particular place and the reasons for this. If they are able to do this they are more likely to find sympathy (or avoid antipathy) and to gain the confidence of people in other areas. For, so frequently, it is the irrational that controls, and to seek explanations among irrational considerations may be a quicker way to the truth than to follow the path of reason. Second, area studies are concerned with a disproportionately large number of seemingly unsophisticated considerations. For example, it is no good shrugging off the influence of geographical environment as something that went out with Semple and Huntington. In a broadcast in 1968, a Trinidadian cabinet minister explained the principal distinguishing fact between him and his English neighbours by saying, "I am a dry season and wet. He is a spring, summer, autumn and winter". Is science anywhere near explaining such a statement—a statement which is of the essence of human geography?

The need for people who are sensitive to the problems of other areas, who are prepared to enter into commitments to them, and who are able to obtain the confidence of their inhabitants, was never greater. At the same time there were never more professional geographers occupied with the interpretation of distant places, the clarification of their problems, and the search after solutions for their difficulties. The older among them probably owe their interest and devotion to the influence of former generations of teachers or to the accident of wartime circumstance. The younger have been deliberately encouraged to develop interests in other lands through the assistance of Treasury funds. Following the report of the Scarbrough Commission in 1947, steps were taken to encourage teaching and research in the field of Oriental, Slavonic, East European, and African Studies. In 1961, a subcommittee, popularly known as the Hayter Committee, reported on the progress made in the university sector since the Scarbrough Commission and put forward additional recommendations. Parallel proposals were put forward for Latin American Studies by the Parry Committee in 1965. Geographers with research interests in these areas are able to apply for the critical financial assistance needed for visits to the Latin American countries and for work in them. For all research in foreign areas, single visits are not enough. Continuing contact is the primary need, and this can be personally demanding as well as financially expensive. It goes without saying that the pursuit of any form of research activity in other areas requires that a working knowledge of the appropriate languages be added to the techniques of the systematist. The same holds increasingly for students.

To conclude, area studies are in some measure necessary to all geographers, but will claim a different degree of allegiance at different stages in the educational programme. In school, the study of other lands remains a basic part of general education and can best be done through geography. At the undergraduate level, students may be given the option of studying areas that appeal to them, but there is good reason for retaining a relatively intensive study of at least one area as a general requirement. At all stages along the line, there ought to be geographers anxious to develop at least some knowledge of this character. To deny it is to lop off a branch of considerable educational value. It is very easy to dismiss such learning as beneath the dignity of the university syllabus and to argue that the more fashionable study of techniques is of more critical concern. Both are needed. An appreciation of the character of locality must be complemented by an understanding of the systematics of location. Otherwise, the lament of Barnabe Googe may again echo in the land:

> Alas that here were Ptholome
> With Compasse, Globe in hande,
> Whose Art should showe me true the Place
> And Clymate where I stand.

At the research level, there is a natural bifurcation of interests among those who look to area studies. It springs from the purpose of the interest. First, there are those for whom an appreciation of the wider geography of particular areas is critical if they wish to practise their systematic training and expertise in them. The contribution of the systematist is likely to be the more effective, in proportion to the breadth and depth of the knowledge that he has of the area where he proposes to employ his skills. Second, there are those for whom area studies are the core of the subject. Enthusiasts who immerse themselves in foreign lands and cultures and who devote a lifetime of interest to their associated phenomena spring even more naturally out of geography than out of history, literature, or languages. In a community which pays more attention to practical and applied considerations, demand for them may dwindle as it does for the services of the classicists. But their choice in favour of such a line of study will be made within their own system of values. In this system, elements of the general will have been retained beside the particular; qualitative experiences beside quantitative assessments; subjective considerations beside objective. Within the hive of geographical studies, those who elect to pursue area studies may add proportionately less to the waxen structure of the comb, but they will continue to contribute a fair quota of honey to fill its cells.

REFERENCES AND FURTHER READING

ARDREY, R. (1967) *The Territorial Imperative*, Collins, London.
DUHAMEL, G. (1931) *Géographie cordiale de l'Europe*, Mercure de France, Paris.
GARLAND, H. (1963) *Crumbling Idols*, Harvard U.P., Cambridge, Mass.
GOOGE, B. (1563) *Eglogs, Epytaphs and Sonettes*.
HUXLEY, A. (1937) *Stories, Essays and Poems*, Dent, London.
LAWRENCE, D. H. (1924) *Studies in Classical American Literature*, Phoenix, London.
READ, H. (1963) *The Contrary Experience*, Faber, London.

Five articles are directly relevant to this chapter:

CLARKE, A. H. (1962) Praemia geographiae: the incidental rewards of a professional career, *Annals of the Association of American Geographers*, **52,** 229–41.
MEAD, W. R. (1963) On adopting other lands, *Geography*, **48,** 241–54.
PRINCE, H. C. (1961–2) The geographical imagination, *Landscape*, **2,** 22–5.
SAUER, C. O. (1956) The education of a geographer, *Annals of the Association of American Geographers*, **44,** 287–99.
WRIGHT, J. K. (1947) Terrae incognitae: the place of imagination in geography, *Annals of Association of American Geographers*, **37,** 1–15.

The enjoyment by a geographer of other places is illustrated by the collection of papers in:

SPATE, O. H. K. (1966) *Let Me Enjoy*, Methuen, London.

A book which illustrated the vicarious manner in which travel literature can enter into the appreciation of the home locality is:

CHRISTIE, J. A. (1965) *Thoreau as World Traveller*, Columbia U.P., New York.

The character and problems of regional geography are brought out in:

MINSHULL, R. (1967) *Regional Geography, Theory and Practice*, Hutchinson, London.

CHAPTER 24

CULTURAL AND HISTORICAL PERSPECTIVE IN AREA STUDIES: THE CASE OF LATIN AMERICA

D. J. ROBINSON

PEOPLES AND PLACES

To an English ear the very term "Latin American" evokes the exotic, and in many ways preconceptions underestimate reality. Contrasts, diversity, complexity—such are the words that readily come to mind when one attempts to describe the peoples and places of this culture area. The range of physical environments is only matched by the differentiation of cultural adaptation, the variability of shapes of phenomena, both natural and human, can only be compared to the equally incredible variety of sizes. Diversity can be seen in all dimensions; longitudinally, latitudinally, and altitudinally regional variations exist in landscapes, economies, and societies. The inhabitants of many such regions have evolved a distinctive vocabulary, used in both written and spoken form, to emphasize peculiar combinations of variation.

The combination of size and shape allows single states to include deserts, rain-soaked, wind-swept forests, and mediterranean gardens. Elsewhere the variation within coastal zones contrasts with the apparent monotony of interior landscapes. Again the observer, like the writer who passes on impressions and judgements, cannot fail to note the tricks which scale plays on the unwary. Just as "cultural spectacles" and the polarization of personal perception affects what one "sees", so too does the speed of travel and the duration of observation. From a jet aircraft flying at 35,000 ft, from a helicopter hovering at 500 ft, from a horse, or from the height of one's own eyes, different things become more or less significant. Monotonous forests become complex, exceedingly varied phenomena; flat grasslands can be seen to be neither flat nor simply grasslands. All depends, of course, on the visibility of these elements of variation, or on the excellence of the observer's vision.

Just as places differ, so too do the peoples and ways of life of Latin America. The range of social and racial types to be found there is, at the present time, as great as may be found in any other continent. There are Chinese,

253

Africans, Negroes, Amerindians, Arabs, and Europeans. While expanding market economies transform the subsistence patterns of the past, conservative peasant farmers clash with revolutionary innovators. The slow pace of life in the "interior" is highlighted by the bustle of the metropolitan city. The aeroplane has made it possible to traverse the continent in a day, yet on the ground, beyond the short ribbons of smooth tarmacadam, locomotion remains necessarily slow.

Contrasts are not only to be seen in Latin America, but also can be seen to be increasing. Awareness of them is enhanced either by their physical juxtaposition, or by the knowledge of their existence. Super-cities are fringed with slums, the rich live close to the poor. Information of all kinds is now spread by radio rather than by the slower, haphazard media of newspapers and books. Technology has put new tools, or weapons, in the hands of those that have power. Soon, in Brazil, the educational satellite may add yet another dimension to diffusion. While news, jurisdiction, and control reach out to previously "remote" areas, people increasingly congregate. In the growing migration from country to town can be seen just one of the many effects of the "revolution of rising expectations". Just as life has changed for many Latin Americans, so also have their attitudes to life. Long-established and accepted order is being challenged. Differences that were previously understandable in terms of the physical gulf that separated societies and cultures, now are only comprehensible in terms of the new gaps that have developed between the young and the old, the rich and the poor, those that govern and the governed, those that make decisions, and those that have no rôle in influencing decision making. New differences between sectors of society have replaced or added to the old ones to make the situation ever more complicated. Whereas, previously, *mañana* or the lottery were the only hopes of fame and fortune, today new paths of progress are being sought.

Though one may find mild amusement in the Latin American method of changing governments, or in their *siesta* habit, or in their apparently "inefficient" attempts to cope with the modern world, or even in their lack of a sense of urgency, in several respects it is difficult not to admire their attitudes. Idealism, be it in presidents or peasants, is an enduring and endearing attribute of Latin Americans. Somehow one feels that they attempt to live whatever life they have to the full. Every man is a potential philosopher, artist, and politician. Though time for action may be desperately short, time is made for talking. Meals remain not simply occasions on which to take in the necessary calories, but are the means to air views and voice opinions, often at length, and potentially on every subject. Latin Americans like to be "involved", and the visitor who wishes to be reinvited to the house that, on his first visit, is offered as a new home, also needs to become "involved" himself. The characteristic initial formality of a host can soon make way for exhuberant informality, friendship, and loyalty.

The problems that face the teacher concerned with increasing his students' awareness of the peoples and places of Latin America are manifold. That there are differences between "them" and "us", between "their" places and "our" places, is undeniable. The task **is** that of deciding upon the best method to describe, to understand, and, if possible, to explain those differences. Can one synthesize the contemporary Latin American scene, and if so, how? What are the best methods of systematizing and rationalizing the apparent processes of change, or of reducing the changing, overlapping patterns to some order? Should one attempt to evoke the "feel" of land and life, and if so, how? These are questions that relate to probably the teaching of any "foreign" area, but the solutions here offered are designed to aid those concerned only with Latin America. In other areas different peoples and places necessarily demand different solutions. The suggestions below also undeniably reflect the author's training and experience. If history and culture appear to obtrude upon the geographical scene it is because it is thought that, at this continental scale, by appreciating what has happened it may be possible for students to better understand what is happening. If spatial analysis is not stressed it is because, at the present time, given the available Latin American data, together with the techniques of the trade, it is thought that time is more profitably spent not statistically describing flows or analysing patterns and the like, but in understanding the origins of some of the many variables. Emphasis is put on the past as a key to understanding the significance of the present; the evolving patterns and processes are followed not by way of systematic threads, but rather are viewed through the more general fabric of cultural development. As such the method has its limitations, it is necessary to generalize, to differentiate between the normal and the abnormal, the sample and the example, but these are problems at the heart of geography itself, and eventually the method can only be judged by its effectiveness or one's own preference. What is clear is the fact that the "foreign" nature of other areas cannot be easily understood by using statistical shorthand. The shape of a state may be more precisely described by geometric expression than by literary approximations, but what should surely be of equal interest to teacher and student alike is why the state is that shape. Similarly, to attempt explanation of current patterns and processes one needs to know something of their provenance.

DESCRIPTION AND EXPLANATION:
THE CONTINENT IN TEXTBOOKS

Many schoolteachers, as well as their pupils, must have noticed the ambiguity of terms used to define major regions within the Americas. Some are political definitions (the United States and Canada) others cultural ("Latin" and "Anglo" America); yet others were based upon structural units (North and South America) and more recently upon distinctive portions or admixtures of culture areas (Middle America). Whereas the physical geography of the Americas was clear-cut, structural geometry had designed, so it appeared, two triangular masses placed one upon the other with an island-studded sea between, human geography had made the picture so untidy. One has parts of the South American mainland more like Anglo America than parts of Anglo America itself; one can find areas in the Caribbean that have shifted their cultural allegiance, so to speak, to make definition dependent upon the time period concerned. One can find in Latin America, areas undeniably more Indian than "Latin" in aspect (Andean America) whilst other areas have curious intermixtures of both Latin and Anglo-American elements. It is clear that the definition of the region is, or rather should be, determined by the purpose to which one wishes to use the region itself.

In the past, since regional differentiation within the major region was usually the subject of study, any ragged cultural edges could be smoothed over within the section dealing with "historical background". Once into the descriptive text the reader could soon be lost in the regional maze. Patterns, processes, peoples, places—all combined to demonstrate with devastating clarity the complexity and the difficulty of understanding the contemporary scene. Having read such texts the student, rather than being inspired to search out the cause of variation within the area, soon lost interest altogether. The region itself became a hold-all from which the teacher selected the more interesting systematic themes to reawaken the flagging interest of his students.

What is here suggested is a more historical and cultural approach to the problem of teaching Latin-American geography. Once the student has been made aware of the diversity, complexity, and heterogeneity of the present scene, he should then be introduced to the two principal stages through which most parts of Latin America have passed, or are presently passing. These are described below as Traditional and Transitional Latin America. The first phase lasted from approximately A.D. 1500 to 1800, the second from the latter date to the present. Whilst the developments in both phases affected the majority of peoples and places of Latin America, they by no means affected all. Exceptions and variations from the general theme of development are the hallmarks of the phases themselves. However, by first "identifying" the establishment of a cultural stage—traditional "Latin" America—and then secondly the modifications that have affected it since new forces began to influence it in the early nineteenth century, it is thought the students will better understand the confused image of contemporary Latin America. In this way the essential cultural unity of the area is emphasized and the student can decide for himself to what extent "Latin" America is becoming less "Latin", and perhaps speculate concerning the future significance of both the area and its name.

TRADITIONAL LATIN AMERICA

Latin-American development has been influenced, some might say determined, by the cultural cargo carried for three centuries from the Iberian peninsula. Wherever Spain and Portugal were the dominant cultural colonists (some might think it is a pity the term "Iberoamerica" could not be used in English literature), a new American tradition was developed (Wagley, 1968). Values, opinions, ideals, views of the world—all were introduced, diffused, and differentiated within the Americas. The individualistic, aristocratic, Catholic, semi-feudalistic cultural traits of southern Europe lay at the base of Iberoamerican colonialism. It was these cultural traits that gave rise to a distinguishable culture area, Latin America. Naturally enough the evolution of the colonial empires provides examples of differential adaptation: in some regions such as Andean America, Amerindian cultures survived or resisted contact and acculturation; elsewhere African Negroes become important racial constituents. Indeed it might be said that the art of adaptation rather than innovation was one of the keys to success for both Spain and Portugal between A.D. 1500 and 1800. In more senses than one the New World provided a new, open frontier denied these two nations in Europe. The characteristic and unifying features of the colonial endeavour were legion. The extended kinship group of the new Iberoamericans soon became a force for stability, for the development of close links between successive generations, for the establishment of élites. The Roman Catholic church was ever-present. First the cross advanced with the sword; next Indo-European barriers were surmounted by missionary linguists; later, secular attitudes to the New World and its inhabitants were formulated in the light of ecclesiastical moral opinion. The church became a firm buttress to all secular institutions imported from the Old World. Land, wealth, and power became equated. Prestige, dignity, and "face" were merely outward expressions of the class-conscious colonial society that evolved in Latin America, and that still is to be found there today, albeit in modified forms. Economic activities and organization reflected the spirit and purpose of the colonies. Mineral "booms" were one key to easy and rapid profit; or, alternatively, it was necessary to introduce and evolve complicated plantation systems. Everywhere could be seen the adaptive colonial hand, be it in accepting aboriginal crops or American methods of cultivation, or the introduction of a well-tried slave-labour system, irrigation methods, or systems of livestock ranching. Most action was deliberate, yet chance too must have played some rôle in the tide of events.

The symbol of civilization was the city. The newly planted towns recreated a little portion of the image of the homeland. What mattered was not the size, or form, or function of the settlement, but its status. The nearer one was to the *plaza*, the farther one was from barbarism. Influenced by winds and tides, by good luck or judgement, the process of colonial rule advanced from first island, and later coastal footholds. Rapidly the new lands were reconnoitred. Appraisals of peoples and places were made and decisions taken. By careful strategy and skilful tactics, slowly a set of new landscapes, economies, and societies were created. Over vast distances and including an incredible range of environments and pre-existing cultures, Spain and Portugal attempted to implement regional planning on a continental scale. Via the bureaucratic machinery culture was allowed to diffuse along historic trails. From Tierra del Fuego to the region of present-day New Mexico the Latin-American tradition affected the land and its peoples. In part, the retreat of the frontier of the unknown was not matched by that of land occupation. Possession and title were what mattered, not occupation and use. It was found easier to congregate Indians and settle Negroes than it was to persuade the Iberian newcomers to farm the land.

Language was yet another unifying force. Though syntax, vocabulary, and pronunciation were gradually, like most other importations, to become differentiated, the most significant fact was the spread, acceptance, and retention of Spanish and Portuguese throughout the continent. Though the Spanish "Indies" and Portuguese "Brazils" were as different from each other as was Spain from Portugal, and though there were regions where "Plantation" or "Indo" America might have been more adequate descriptions, by the late eighteenth century a new culture area had been created in the New World. In north-west Europe it went by the name of Latin America.

TRANSITIONAL LATIN AMERICA

Despite at least the Spanish Crown's attempts to ensure otherwise, traditional Latin America was never entirely free from the effects of contact with, and changes taking place within, non-Latin culture areas. Indeed, as has been seen elsewhere in the world, it was only when the traditional order had been established, that new forces, both internal and external, became active. Though the origins of the new forces were closely linked to the growth of industrial Europe, their impact had to await the severing of colonial ties with both Spain and Portugal in a series of revolutions during the early nineteenth century. Only then did economic, social, and personal adjustments become necessary or even possible. From the beginning in transitional Latin America political change was the precursor of all other change. It was now the privilege of each of the new Latin American republics to appraise their situation, to judge the merits of the "Latin" American tradition, and, where it was thought necessary, or found possible, to effect change within that order. This each republic has been engaged upon from the early nineteenth century to the present, 150 or so years of modification.

Since in Europe, Anglo-America, and elsewhere the agricultural age had made way for industrialization, the newly sensitivized culture area of Latin America was almost bound to feel its impact in some form or other. Technological innovations, which had previously been excluded, now entered into agriculture, mining, and urban life. Appraisals and reappraisals were now possible. Areas of the continent formerly considered marginal, became new heartlands of economic activity. In the temperate grasslands of Argentina, for instance, cattle ranching changed from a prestige pastime to a capitalistic industry closely integrated with the meat-packing plants and the Plate ports that led to north-west Europe. New breeds of cattle and varieties of grasses accompanied non-Latin colonists, investors, and entrepreneurs. Tramways in the towns and railways in the countryside became emblems of modernity. The flow of meat and men to and from Europe spread knowledge of the new lands, empty, inviting, and available. Facts and fancies lured English, Welsh, Italians, and others to seek their eldorados. The traditional order had to change. Those fortunate enough to own land either farmed in the new manner or sold out to build their *palacios* in the cosmopolitan capital. In the countryside the frontier of settlement and effective control advanced southwards so rapidly as to leave little historical record. The tempo of change rapidly accelerated. The peasant society of the *littoral* was dragooned to ward off that impertinent enemy of progress, the Indian: Argentine Araucania was rapidly transformed with landscaped gardens *al estilo inglés*, set amidst wire-fenced fields.

The story of temperate Argentina was re-enacted with different scenarios and characters in many other parts of Latin America. Whilst the new politicians were busy trying to mould their philosophical ideologies into practical instruments of government, economies, landscapes, and ways of life were changing. In Chile the exploitation of mineral deposits in the north and timber resources in the south produced new settlements, new élites, and an accentuation of regional distinctiveness. Nitrate kings became the Chilean equivalent of the Argentine cattle barons. The traditional economic, political, and social structures strained under the impact of new capital, ideas, techniques, labour, and a new, more insidious, form of colonialism. The new republics began to realize the disadvantages of their inheritance, and to readjust accordingly. With markets, prices and production open to world competition and comparison, when premiums were set upon productivity and efficiency measured by new standards, Latin America fell into the ranks of the "underdeveloped" areas of the world. Cultural identity became equated with economic problems, progress, and potential.

With the new changes came increasing concentration and localization of power and wealth. Sections of the traditional societies reverted to subsistence levels. In Andean America beyond the tin mines, railways, and towns, Indians appeared even more isolated from the modern world. The benefits of political independence and economic development appeared to be only the perquisites of the new aristocracy. The variability in patterns and rates of adjustment within Latin America makes generalizations almost impossible, yet at the same time it raises many questions which require answers. Three processes were to affect the pace and location of adjustment throughout Latin America; they were the rapid growth of population caused by a sharp decline in mortality rates, the associated process of urbanization, and, more recently, the impact of commercial markets on subsistence agriculture. These three processes, like each of the many factors affecting the traditional order, have produced, and are producing, variable results. Almost every change involved conflict, and each nation has had to decide whether evolution or revolution is the quickest, most beneficial, or most desirable method of allowing or causing change to take place. Technology is, in a way, outstripping institutional adjustment to new problems and potentials. Latin America today can, however, be usefully regarded as an area that is in a phase of transition. Just as it took three centuries for the New World to be modelled on the lines of the Old within the Iberian tradition, so it may take many more decades before that Latin American tradition can adjust itself to the new forces emanating from Anglo America, Europe, and elsewhere.

The present complexities and diversity can thus be more easily understood. They represent the patterns resulting from modifications made to older structures and systems evolved under the peculiar conditions of colonial rule. It is suggested that only by giving due weight to the process of historical development can either teacher or student of Latin American geography become aware of its relevance to the contemporary scene.

THE LITERATURE OF LAND AND LIFE

It was perhaps to be expected that the most eloquent appraisal of the transitional situation in which Latin Americans found themselves after the first decades of the nineteenth century, should have come from the literary intelligentsia. Theirs was the task of searching the traditional past for moral values, national origins, and ways of life that could form an integral part of the new societies that were forming. Alienated from their Latin heritage by force of social, economic, and political circumstances, they sought roots in the land, the Negro, and aboriginal Indian cultures. In the eyes of the urban, or urbanized, poet or novelist the land possessed special qualities. Unlike the transplanted, artificial world of the city, it was perceived as a hostile realm of nature. Beyond the fringe of urban life lay not an Arcadia but a land that, through its character, so it was thought, would mould the people who inhabited it. Rojas (1916) in Argentina, argued that the *fuerza territorial* (the pull of the land) would make the flood of new immigrants Argentinian; soil and seasonal rigours of the climate would be catalysts of adaptation. Similarly in the poetry of Gabriela Mistral (1922) and Juana de Ibarbourou (1960) the formative qualities of the land were central themes. Perhaps it was in the novel and the short story, however, that this notion achieved its most notable protagonists. In the writings of Güiraldes (1926), Quiroga (1918 and 1921), and Gallegos (1929, 1934, and 1935), the ephemerality of urban civilization was contrasted with the durability of the qualities that nature bestowed upon man. These writers related peoples to places.

Indigenismo (Indianism) became another source of literary inspiration. It served a dual purpose for society in so far as it first drew attention to the plight of the underprivileged masses, and secondly it became possible, through a study of indigenous origins, to suggest alternative values to those of European origin. In Mexico this literary movement was closely associated with "muralist" expressions of similar themes (Charlot, 1963). The gulf that existed between White and Indian society in the early part of the present century is dramatically portrayed in a series of novels, several of whose authors were trained anthropologists (Pozas, 1952; López y Fuentes, 1935; Magdaleno, 1937). Indian attitudes to the "sacredness" of land and its exploitation, their values and psychological frames of reference were also explored (Asturias, 1946 and 1949; Castellanos, 1957). The political relevance of Indian problems (Alegría, 1941) and their potential threat to social order (Arguedas, 1958) provided constant sources for literary inspiration and interpretation from the 1920's.

Where it was less reasonable to turn to indigenous cultures to seek out national roots, the Negro provided a substitute. In the Caribbean islands this was a dominant theme (Coulthard, 1962). Life on the old slave plantations and the contrasts between urban and rural living were brilliantly recreated. In Brazil, where one might have expected to have found a dynamic Negro-centred literature, a new product, the "Brazilian", had been created. As Freyre (1933 and 1945) and his school of social historians had so dramatically demonstrated, in Brazil Indian, White, and Negro cultures had been fused by miscegenation into a new hybrid culture, that itself could be identified with Brazilian nationalism. In Brazil, too, regionalism, was a key theme of literature. Many authors described the origins of, and developments within, the "Brazils" (Lins do Rego, 1935; Ramos, 1938; Queiros, 1931). If one is seeking the "feel" of the *sertão* (the Brazilian outback) or the *nordeste* (the north-east sugar-cane area), then the regional novelists provide excellent indigenous appreciations. Likewise in Argentina a rich descriptive regional literature developed from the late nineteenth century. The epic story of the changing *pampa* and its folk-figure, the *gaucho*, make fascinating reading (Hernândez, 1872; Lugones, 1917).

Even with such a selective and small sample of the literary product of Latin America it will be appreciated that literature presents a wealth of illustrations for students concerned with the many geographical aspects of land and life. In the writings of Latin Americans one can see something of their view of their culture, and environment, often a significantly different view to that of European visitors or residents. At the same time one can identify in much of the literature the new forces of transition in action, be it the decline of the sugar-plantation, the railway "boom", or the breakdown of social structures. As the writers live, or lived, in the phase of transition, so one can benefit from their opinions, perceptions, and judgements on the traditional past and its relevance for contemporary life.

THE MESSAGE OF MUSIC

The language barrier, still a formidable obstacle for many students of Latin America, has undoubtedly prevented a wider appreciation of Latin American literature. Yet, as far as its music is concerned, most would feel capable of recognizing Latin American dance rhythms such as the rumba, samba, or bossa nova. Despite this superficial (and partly misleading) acquaintance, however, many students might be hard pressed to see the geographical relevance of Latin American music. Its virtual neglect by geographers is all the more surprising, considering the quantity of information available, gained from the studies of ethnographers, anthropologists, musicologists, and the several distinguished schools of folklore in the area. The origins and diffusion of types of instruments, of peculiar musical form, and of rhythms, provide one with valuable insight into the culture history of the area. As will be shown below, by reference to a selection of published records that can be recommended to any teacher of Latin American geography, much of the contemporary folk music of the continent reflects images of the past.

In several areas of Latin America there exist groups of Indians whose cultures appear to have been little affected by Europeans. In their songs and dances one hears echoes of the pre-Hispanic past. Tribal ceremonies, special food preparation, imitative animal hunting cries—all are recorded in their music (Venezuelan Folk Music). In other areas in addition to their range of primitive bark, wood, and bone instruments, one can hear the European-style skin side-drum (*caja*), introduced in this case probably by Franciscan missionaries in the last century (Música Guajira).

With the Spanish came the guitar and the harp and in the lyrics of the *polos* and cadences of the *galerones* one can hear influences of Andalucian Spain in the folk music of eastern Venezuela (Venezuelan Folk Music). In the popular four-stringed *quatro*, too, one can see a derivative of one of the imported sixteenth-century Spanish guitars. Just as literature reflects the significance of Negroes in certain regions in the past, so too, does music. Instruments such as the *quitiplas* (stamping tubes) and *carangano* (musical bow) derive directly from Africa; the *fulia* songs sung at the *velorios* (wakes), accompanied by various *tambores* (drums), remind one of the plantation era. The plainlands also have their distinctive music, characterized by cattle-herding lullabies and driving songs, similar in sentiment to those of the grasslands of the United States (Venezuelan Folk Music; Contrapunto). Whereas in Venezuela it was the struggle of man against the tropical forest and savannas that produced some of the finest folk music (Séptimo Paralelo), in Argentina it was the conquest of the *pampa* in the 1880's that provided sad songs of the break-up of families as the menfolk are sent south to garrison the frontier forts (Martín Fierro).

In the Caribbean islands and Brazil colourful rhythms introduced from Africa and fused with creole melodies gave rise to the rumba, samba, and bossa nova, each of which, in its true form, is surprisingly different from its popular export variety (Batucada). In the Andean highlands the diffusion of musical instruments along historic trails such as that leading south from Potosí to northwestern Argentina has left easily recognizable affiliations between the music of the two areas. The *charango* (small type of mandolin) and *quena* (small recorder) are now to be found well outside their true home, the high mountain plateaux of Bolivia (Taquirari).

The influx of immigrants into Latin America during the latter half of the nineteenth century can also be traced through their effect on creole music. In Argentina, for example, cosmopolitan Buenos Aires provided a musical melting-pot. In the port area of "La Boca" there arose, as a result of Italian immigrants mixing with creole musicians and Negro sailors, a new dance and music. First it was known as the *tambó*, only later being called by its better-known name, the *tango*. The increasing popularity and diffusion of the new style of music, from the dockside bars to the cafés, then into the fashionable salons of higher society, can be traced by means of the form of the arrangements, as well as of the type of solo instruments employed. First the violin, then the piano, later the *bandoleón* (type of concertina) and finally full orchestras played the new musical symbol of Argentina's modernity (El Tango). Eventually the *tango* was exported to the world by one of that country's most famous sons, the singer Carlos Gardel (Carlos Gardel).

CONCLUSION

In landscapes, cultures, and societies, just as in music and literature, the past is ever present. It may be the past of a decade before, or a century, or many centuries. There can also be little doubt that the contemporary scene in Latin America is characterized by change. Indeed one recent author emphasized the inadequacies of "traditional ways of studying a continent the size of Latin America, in such a state of change" (Cole, 1965, p. 441). If change is the focus of attention at present, then it might be better to concentrate on that process rather than to be lured into the morass of regionalization. Similarly, as has been suggested above, present processes of change can only be fully understood in their Latin American context by putting them in the necessary historical perspective. Only by recognizing the present as a temporary stage in the phase of transition from what was traditional Latin America can the student fully understand the contemporary scene. If he finds it possible to understand the process of Latin American cultural and historical evolution, it might also help him to understand that of his own culture area, to realize that Latin America is but a portion of the inhabited world that has special lessons to teach us. If by looking at such a "foreign" area the students' sensibility to what is termed "cultural relativism" can be increased, then the task of teaching will not be in vain.

REFERENCES AND FURTHER READING

PEOPLE AND PLACES

The following short selection of books gives some idea of the variety of topics and methods of presentation:

BURRI, R. (1968) *The Gaucho*, Macdonald, London.
CUNHA, E. DA (1947) *Revolt in the Backlands*, Gollancz, London.
RICH, J. L. (1942) *The Face of South America: an aerial traverse*, American Geographical Society, New York.
SIMPSON, L. B. (1967) *Many Mexicos*, U. of California P., Berkeley.

DESCRIPTION AND EXPLANATION

For an outstandingly lucid text that skilfully blends description and explanation see:

WEST, R. C. and AUGELLI, J. P. (1966) *Middle America: its lands and peoples*, Prentice-Hall, Englewood Cliffs.

For sacrifice of explanation at the altar of regionalization see:

JAMES, P. E. (1959) *Latin America*, Cassell, London,

and for a clear statement concerning the general problems of area studies see:

STEWARD, J. H. (1950) *Area Research: theory and practice*, Bulletin, **63,** Social Science Research Council, New York.

TRADITIONAL LATIN AMERICA

For further information concerning cultural relativism, cultural components, and areas, see:

FOSTER, G. M. (1962) *Traditional Cultures: and the impact of technological change*, Harper, New York.

HEATH, D. B. and ADAMS, R. N. (1965) *Contemporary Cultures and Societies of Latin America*, Random House, New York.

WAGNER, P. L. and MIKESELL, M. W. (1965) *Readings in Cultural Geography*, U. of Chicago P., Chicago.

WAGLEY, C. (1968) *The Latin American Tradition*, Columbia U.P., New York.

The development of the traditional order is dealt with in:

AZEVEDO, F. DE (1950) *Brazilian Culture*, translated by Crawford, W. R., Macmillan, London.

BAILEY, H. M. and NATASIR, A. P. (1960) *Latin America: the development of its civilization*, Prentice-Hall, Englewood Cliffs.

CLISSOLD, S. (1965) *Latin America: a cultural outline*, Hutchinson, London.

DIFFIE, B. W. (1967) *Latin American Civilization: the colonial period*, Octagon Books, New York.

WORCESTER, D. E. and SCHAEFFER, W. G. (1956) *The Growth and Culture of Latin America*, Oxford U.P., London.

TRANSITIONAL LATIN AMERICA

An excellent handbook with valuable bibliographies is:

VÉLIZ, C. (Ed.) (1968) *Latin America and the Caribbean. A Handbook*, Blond, London.

The genesis of transition is outlined in:

FURTADO, C. (1963) *Economic Growth of Brazil*, U. of California P., Berkeley.

MANNERS, R. A. (1964) *Process and Pattern in Culture*, Aldine P., Chicago.

WAGLEY, C. (1963) *An Introduction to Brazil*, Columbia U.P., New York.

The progress of transition can be followed in:

COLE, J. P. (1965) *Latin America: an economic and social geography*, Butterworth, London.

D'ANTONIO, W. V. and PIKE, F. B. (Eds.) (1964) *Religion, Revolution and Reform: new forces for change in Latin America*, Praeger, New York.

HAUSER, P. (Ed.) (1961) *Urbanization in Latin America*, UNESCO, Paris.

JOHNSON, J. J. (Ed.) (1963) *Continuity and Change in Latin America*, Stanford U.P., Stanford.

SILVERT, K. H. (1961) *The Conflict Society: reaction and revolution in Latin America*, The Hauser P., New Orleans.

TEPASKE, J. T. and FISHER, S. N. (1964) *Explosive Forces in Latin America*, Ohio State U.P., Columbus.

VÉLIZ, C. (Ed.) (1965) *Obstacles to Change in Latin America*, Oxford U.P., London.

VÉLIZ, C. (Ed.) (1967) *The Politics of Conformity in Latin America*, Oxford U.P., London.

More specific studies are:

AZEVEDO, T. (1963) *Social Change in Brazil*, U. of Florida P., Gainsville.

GRAHAM, R. (1968) *Britain and the Onset of Modernization in Brazil*, **1850–1914**, Cambridge V.P., London.

LEWIS, O. (1961) *The Children of Sanchez*, Random House, New York.

NICHOLSON, I. (1966) *The X in Mexico: growth within tradition*, Faber, London.

THE LITERATURE OF LAND AND LIFE

An outstanding survey of the Modernist period, with excellent bibliographies is:

FRANCO, J. (1967) *The Modern Culture of Latin America*, Pall Mall P., London.

Three useful anthologies are:

ARCINIEGAS, G. (1947) *The Green Continent*, Editions Poetry, London.
COHEN, J. M. (Ed.) (1967) *Latin American Writing Today*, Penguin, Harmondsworth.
Encounter (September 1965) *Rediscovering Latin America*, special publication, London.

For a wider view see:

ENGLEKIRK, J. E. (Ed.) (1965) *Outline History of Spanish American Literature*, Appleton-Century-Crofts, New York.

In the following list of works referred to in the text, where possible English versions have been cited, the dates of their publication being placed with the publisher:

THE LAND

ALERGÍA, C. (1941) *El mundo es anchoy ajeno*; published in English as *Broad and Alien is the World*, translated by Onis, H. de, Nicholson & Watson, London, 1942.
ARGUEDAS, J. M. (1958) *Los ríos profundos*, Losada, Buenos Aires.
GALLEGOS, R. (1929–35) *Doña Barbara* (1929); *Cantaclaro* (1934); *Canaima* (1935); are all published in *Obras completas*, Aguilar, Madrid, 1958.
GÜIRALDES, R. (1926) *Don Segundo Sombra*, in *Obras completas*, Emece, Buenos Aires.
IBARBOUROU, J. DE (1960) *Obras completas*, Aguilar, Madrid.
MISTRAL, G. (pseudonym of Lucila Godoy Alcayaga), (1922) *Desolación*, in *Poesías completas*, Aguilar, Madrid.
QUIROGA, H. (1918) *Cuentos de la selva para niños*, Losada, Buenos Aires.
ROJAS, R. (1916) *Blasón de plata*, Libería La Facultad de Juan Roldán, Buenos Aires.

THE INDIAN

ASTURIAS, M. A., (1946) *El Señor Presidente*; published in English as *The President*, translated by Partridge, F., Gollancz, London, 1963.
ASTURIAS, M. A. (1949) *Hombres de maíz*, in *Obras escogidas*, vol. 1, Aguilar, Madrid, 1955.
CASTELLANOS, R. (1957) *Balún Canán*; published in English as *The Nine Guardians*, Gollancz, London, 1959.
CHARLOT, J. (1963) *The Mexican Mural Renaissance, 1920–25*, Yale U.P., New Haven.
LÓPEZ Y FUENTES, G. (1935) *El Indio*; published in English as *They That Reap*, translated by Brenner, A., Harrap, London, 1937.
MAGDELANO, M. (1937) *El resplandor*; published in English as *The Sunburst*, translated by Brenner, A., Drummond, London, 1945.
POZAS, R. (1952) *Juan Pérez Jolote*, Fondo de cultura económica, Mexico.

THE NEGRO

COULTHARD, G. R. (1962) *Race and Colour in Caribbean Literature*, Oxford U.P., London.
FREYRE, G. (1933) *Casa Grande e Senzala*; published in English as *The Masters and the Slaves*, translated by Putnam, S., Knopf, New York, 1956.
FREYRE, G. (1945) *Brazil: an interpretation*, Knopf, New York.
HERNANDEZ, J. (1872) *Martín Fierro*, Losada, Buenos Aires, 1959.
LINS DO REGO, J. (1935) *O Moleque Ricardo*, José Olympio, Rio de Janeiro.
LUGONES, L. (1917) *El libro de los paisajes*, in *Obras poéticas completas*, Aguilar, Madrid, 1959.
QUEIRÓS, R. DE (1931) *O Quinze*, in *Três romances*, José Olympia, Rio de Janeiro.
RAMOS, G. (1938) *Vidas Sêcas*, Portugalía, Lisbon.

THE MESSAGE OF MUSIC

A useful short outline of Latin American music, with references, is to be found in the handbook edited by C. Véliz (1968, pp. 814–819), listed above. Examples of folk music available on records cited in the text are:

Batucada, Escola de Samba da Cidade; Philips, P.630.458.L, Rio de Janeiro, Brazil.

Carlos Gardel, Odeon—LDS—762, Vol. 42, Buenos Aires, Argentina.

Contrapunto, Polydor, Mono 010, Caracas, 1963. Arrangements by R. Suarez.

El Tango, Historia de la Orquesta Típica: el tango en su evolución. Difusión Musical, 70.032, Buenos Aires, 1966. Musical direction by Argentino Galvan, explanatory text by Luís A. Sierra.

Martín Fierro, Philips (Argentina) Selección Estelar, 82204 PL, Buenos Aires, 1968. Sound track from the film of the same name, directed by Leopoldo Torre Nilsson, music by Ariel Ramírez.

Música Guajira, Volume I of *Música Indígena Venezolana,* published by the Indigenist Commission, Ministry of Interior and Justice, Caracas, Venezuela, 1963, and directed by F. Carreño.

Séptimo Paralelo, RCA—Victor, PML—1001, Caracas, 1962. Sound track from the film of the same name, directed by Elia Marcelli, musical arrangements by Silvio Estrada.

Taquirari, Philips (Argentina) 82180 PL Mono. Played by J. Torres.

Venezuelan Folk Music, Volume X of The Columbia World Library of Folk and Primitive Music. Edited by J. Liscano and A. Lomax.

CHAPTER 25

LIFTING THE IRON CURTAIN

R. A. FRENCH

THE political Iron Curtain, which came down across Europe after the Second World War, was sadly, if inevitably, reflected in geography. The impossibility of visiting the Soviet Union and the countries of eastern Europe, the absence of any textbooks written after 1939, indeed the almost total lack of information, made the teaching of Soviet and east-European geography an almost hopeless task. For many years even such basic items of information as the total population of the Soviet Union were jealously guarded state secrets. The handful of western geographers interested in the area felt themselves to be detectives, trying to solve a mystery without clues, and more than one gave up in despair. In all but a very few schools the vast communist world became *terra incognita*. It could hardly have been otherwise, when the teacher had nothing whatsoever to rely on, other than a couple of pre-war textbooks and some highly tendentious Soviet publications in English.

THE GROWING AVAILABILITY OF INFORMATION

This regrettable situation persisted with little change until the end of the 1950's. Then the slow, partial thaw, which set in after Stalin's death in 1953, began to have its effects. The 1960's have seen a huge increase in the availability of geographical material at all levels. No longer is there any need, or any excuse, for the neglect of such a vitally significant area of the world. In most east-European countries one may travel freely wherever the Soviet army is not in operation. In the Soviet Union travel is still greatly restricted, but scores of towns in all parts of the country are open to foreigners, who may, if they wish, journey by train from Brest on the Polish frontier to Nakhodka on the Sea of Japan. Certain Soviet ports are visited by educational cruises. Above all, information, if not freely available, is far more abundant than ever before. Yet even now very few schools provide teaching on the Soviet Union or eastern Europe, although most G.C.E. boards offer the option of doing so.

No one would dispute the importance of the region, not only geographically, but also in our day-to-day lives. One suspects that this neglect is primarily due to lack of knowledge in the schools about the newly available source materials.

The lifting of the information barriers has resulted in the publication during the 1960's of a number of textbooks on Soviet and east-European geography, intended for various levels from the junior secondary school to university. A selection is given in the references. Naturally these vary very greatly in quality and usefulness, but now the teacher does have a choice and can find a reasonably helpful and up-to-date textbook to form the basis of a course at most levels. Moreover, a number of articles are published from time to time in geographical and other journals, notably the *Annals of the Association of American Geographers*. Several very useful and well illustrated articles have appeared in *The Geographical Magazine*.

MATERIAL ON THE SOVIET UNION

For the teacher who wishes to look beyond the textbook, a far wider range of sources is currently accessible. The study of geography in the Soviet Union is highly developed and a host of geographers generate a formidable volume of publication, most of which can be obtained outside the country. Few geography teachers have the command of Russian necessary to use this mass of material, but more and more is becoming available in translation. Of outstanding value is the journal *Soviet Geography: Review and Translation*, edited by Theodore Shabad and published ten times a year since 1960 by the American Geographical Society. This comprises translations of articles from Soviet geographical journals. The emphasis is rather on articles dealing with methodology and concepts and with economic geography, although all aspects of the subject are covered to some extent. This indispensable journal not only gives a broad picture of work proceeding in the Soviet Union, but also provides many regional cases and examples, which the teacher may find useful. In addition to the articles in translation, *Soviet Geography* contains notes culled from the Soviet press reporting the latest developments in the geography and economy of the country.

An increasing number of basic works of reference in Russian have been translated into English. Most of these are rather advanced but they provide an invaluable source for the teacher constructing his own course. Deserving of particular commendation to the teacher are Borisov's work on the climates of the Soviet Union, Berg on the natural regions of the country, Suslov on the physical geography of Asiatic Russia, and Saushkin's study of Moscow. These and other translations are given in the reference section. Polish and Hungarian geographers themselves publish much useful work in English.

The teacher need not feel restricted to works in translation. To acquire an adequate reading knowledge of Russian may be difficult, but the ability to transliterate can be achieved with a few hours work. This ability at once opens up the rich treasure house of geographical material in Soviet atlases and statistical handbooks. Many Russian geographical words, once transliterated, are readily recognizable* and familiarity with the Cyrillic alphabet makes it possible to use a dictionary to translate the headings of maps and tables.

In the Soviet Union, topographical maps on a scale larger than 1:500,000 are classified as secret. They are therefore totally inaccessible to the foreigner, and accessible even to Soviet geographers only on a need-to-know basis. No doubt as a consequence, the production of highly detailed, small-scale maps has been developed to a striking degree. Nowhere is this better seen than in the series of regional atlases of individual republics and provinces (Fig. 23). These contain maps of all aspects of physical, economic, and human geography in very considerable detail, including topics rarely mapped on a national scale, such as geomorphology and tectonics. Of great interest and use in teaching are special maps of local features—land use of specific farms in the *Kustanay Province Atlas*, the coal deposits and mines of Vorkuta in the *Komi Republic Atlas*, the Bratsk Reservoir in the *Irkutsk Province Atlas*, for example. In addition to the atlases, the Soviet Union publishes many wall maps for teaching, showing nation-wide distributions of physical and economic features and general economic features of individual areas.

Economic statistics are still very scarce in the Soviet Union, but some figures are now published, notably in the annual handbook *Narodnoye Khozyaystvo SSSR* (*The National Economy of the U.S.S.R.*). This gives data on population, production of the principal branches of the economy, transport, and social services. Another statistical yearbook deals with foreign trade. From time to time handbooks are published giving figures relating to particular branches of the economy—industry in general, the coal industry, the timber industry, agriculture, livestock. Other handbooks refer to individual republics or provinces. The figures provided are very far from complete; many economic activities are never covered, for example non-ferrous mining, shipbuilding, the aircraft industry. Frequently figures are not broken down for territorial units less than a Union Republic. At times, the statistics may be misleading without further information, which is usually not given. Nevertheless there is no reason to doubt that they are meaningful and they permit the compilation of a vastly more detailed and accurate picture of the country's economic and human geography. The handbooks also provide a useful source for school projects and exercises.

---

* For example, a random selection: okean, geomorfologiya, tektonika, platforma, plato, vulkan, geologicheskaya karta, klimat, temperatura, tsirkulatsiya atmosfery, radiatsionnyy balans, ekonomicheskaya geografiya, metallurgiya, atomnaya electricheskaya stantsiya.

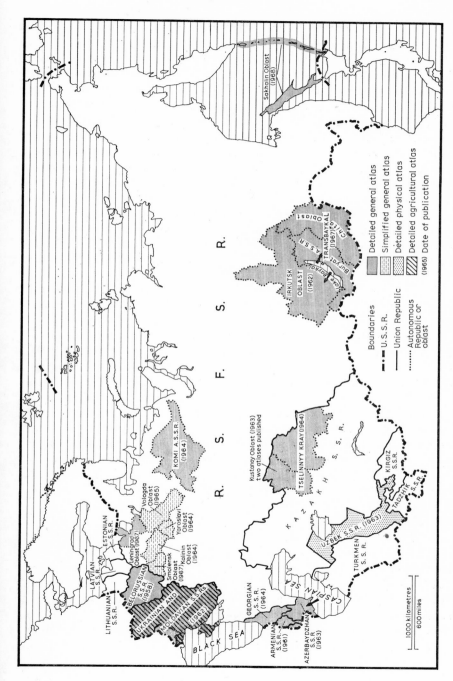

Fig. 23. Areas of the Soviet Union covered by published regional atlases. A number of further atlases are in preparation.

THE STATE AS A GEOGRAPHICAL FACTOR

In dealing with these sources and in teaching the geography of the Soviet Union, it must always be remembered that the country has a command economy. It is therefore as important for the student as for the teacher to have at least an outline conception of how such an economy works and how it affects geography. The extremely important rôle played by the state in contemporary geography has been underlined more than once in earlier chapters, particularly those on applied geography. In the Soviet economy the state is paramount, the only source of investment and the only significant decision-maker, and one may see more clearly than elsewhere the widespread geographical consequences of state policy. If in Great Britain the Barlow Report, the various Town and Country Planning Acts, the Buchanan Report, the Ministry of Agriculture decisions on subsidies are all of vital relevance in our modern geography, how much more so are the plans drawn up by GOSPLAN for the whole Soviet economy and the decisions of the Presidium and Central Committee of the Communist Party on priorities of investment and development.

The power of the Soviet government to control prices and wages, the nature and volume of production at all stages—primary production, transport, processing and manufacturing, retail distribution and consumption—means that the effects of the working of normal economic laws can be modified, or even suspended. The decision in the First Five-year Plan to link the Ural iron ore deposits and the Kuzbass coalfield in a 1200-mile interchange of ore and coke was made "economic" by fixing artificially low freight rates for these materials on the relevant section of the Trans-Siberian Railway. The location of the Komsomolsk steel works in the far east, far from iron ore or coal supplies, away from any market or source of pig iron or scrap and without any pre-existing communications other than the river Amur, is explicable only as an arbitrary planners' decision, taken in the light of prevailing theories of regional development. A number of metallurgical plants established during the Soviet era are very high-cost producers and the reasons for their locations must be sought not in geographical factors, but rather in political decisions and the doctrines which underlie them.

In human geography the state has played a no less significant part. A concomitant of collectivized agriculture has been the gradual replacement of dispersed settlement by nucleated villages. At times the Soviet government has exercised rigid control over population movement. In the Stalinist period, workers were in large measure tied to their jobs, by regulations which Khrushchev subsequently abolished. The coal-mines of Vorkuta, located in a region of permafrost north of the Arctic Circle and many hundreds of miles from any market, could be opened up because labour was an insignificant cost factor, consisting as it did of German prisoners of war and political prisoners. Even today, with most concentration camps emptied and relatively free movement of labour, the growth of cities like Moscow and Leningrad is kept artificially low by the requirement of a police permit to reside in them.

Not only is the hand of the state seen everywhere, but the Soviet love of the large-scale operation, allied to the size of the Soviet Union, has meant that state-sponsored developments have affected enormous areas in relatively short spans of time. Khrushchev's Virgin and Idle Lands campaign brought about the ploughing up of more than 100 million acres within 6 years, with the accompanying foundation of new settlements and the building of new lines of communication. The policy of utilizing the great hydroelectric potential of Soviet rivers has caused the construction of many giant barrages and the drowning of tens of thousands of square miles. Current plans of reclamation envisage the draining of millions of acres of swamp and the irrigation of millions of acres of arid land.

It is therefore of the utmost importance for the teacher to keep abreast of developments in Soviet economic policy. The "Party Line" has made frequent changes of direction. Under Stalin, control of industry was highly centralized in ministries in Moscow and the country was divided for planning purposes into a few, very large regions. The basic principle was that each region should be as self-sufficient as possible. The Khrushchev period saw a total reversal of policy, with central ministries abolished and replaced by local economic councils. The planning regions were more numerous and smaller, based on regional specialization of production. Brezhnev and Kosygin have once more reversed policy, dismantling the Khrushchev apparatus of control and his planning regions and re-establishing the central industrial ministries.

Such changes in policy, with their consequent changes in economic geography, take place rapidly. The economic growth of the country is still relatively rapid. In the 1940's, Baku produced 80 per cent of Soviet petroleum; in the 1950's and 1960's the Volga–Urals field provided 80 per cent of a far greater production. The huge discoveries in western Siberia and Kazakhstan in the later 1960's mean that a new pattern of production is emerging for the 1970's. Even the best of textbooks is out of date almost as soon as it appears. For the teacher no source of information is more important than the daily and weekly press. What Kosygin says today may be the economic geography of tomorrow. Other than the basic physical features there is no aspect of present Soviet geography, which does not in some way reflect state decision or influence. A geographer must inevitably, therefore, take this factor into consideration as carefully and analytically as he would the climatic régime and relief. But given such an approach and given the sources now available, there is no longer any reasons why sixth-form students and undergraduates should not study the Soviet Union with profit and success.

REFERENCES AND FURTHER READING

THE GROWING AVAILABILITY OF INFORMATION

General regional textbooks, suitable for use by sixth-form students and undergraduates are:

COLE, J. P. (1967) *Geography of the U.S.S.R.*, Penguin, Harmondsworth.
COLE, J. P. and GERMAN, F. C. (1961) *A Geography of the U.S.S.R.*, Butterworth, London.
DEWDNEY, J. C. (1965) *A Geography of the Soviet Union*, Pergamon, Oxford.
GREGORY, J. S. (1968) *Russian Land Soviet People*, Harrap, London.
HAMILTON, F. E. I. (1968) *Yugoslavia: patterns of economic activity*, Bell, London.
HOOSON, D. J. M. (1966) *The Soviet Union*, U. of London P., London.
LYDOLPH, P. E. (1964) *Geography of the U.S.S.R.*, Wiley, New York.
MELLOR, R. E. H. (1964) *Geography of the U.S.S.R.*, Macmillan, London.
OSBORNE, R. H. (1967) *East-Central Europe*, Chatto & Windus, London.

MATERIAL ON THE SOVIET UNION

The following are basic works of reference, which have been translated from the Russian:

BERG, L. S. (1950) *Natural Regions of the U.S.S.R.*, Macmillan, New York.
BORISOV, A. A. (1965) *The Climates of the U.S.S.R.*, Oliver & Boyd, Edinburgh.
GERASIMOV, I. P. et. al (1962) *Soviet Geography: accomplishments and tasks*, American Geographical Society, New York.
LYASHCHENKO, P. I. (1949) *History of the National Economy of Russia to the 1917 Revolution*, Macmillan, New York.
NALIVKIN, D. V. (1950) *The Geology of the U.S.S.R.*, Pergamon, Oxford.
SAUSHKIN, YU. G. (1967) *Moscow*, Foreign Languages Publishing House, Moscow.
Soils of the U.S.S.R. (1961) 2 vols., I.P.S.T., Jerusalem.
SUSLOV, S. P. (1961) *Physical Geography of Asiatic Russia*, Freeman, San Francisco.
TSEPLAYEV, V. P. (1965) *The Forests of the U.S.S.R.*, I.P.S.T., Jerusalem.

The following national atlases of the Soviet Union have been published to date:

Oxford Regional Economic Atlas. The U.S.S.R. and Eastern Europe (1956), Clarendon P. Oxford.
The World Atlas (1967) 2nd edn., G.U.G.K., Moscow. (In English, with a large section of excellent physical maps of the Soviet Union.)
Atlas S.S.S.R. (*Atlas of the U.S.S.R.*) (1962), G.U.G.K., Moscow.
Atlas Razvitiya Khozyaystva i Kul'tury S.S.S.R. (*Atlas of the Development of the Economy and Culture of the U.S.S.R.*) (1967). G.U.G.K., Moscow.
Atlas Sel'skogo Khozyaystva SSSR (*Atlas of agriculture in the U.S.S.R.,*) (1960) G.U.G.K., Moscow.

A full bibliography of Soviet statistical handbooks up to the end of 1965 is given in:

KASER, M. C. (1966) 1956–65, Soviet statistical abstracts, *St. Antony's Papers*, **19,** 134–55.

To those listed may be added:

Strana Sovetov za 50 let (*The Land of the Soviets over 50 years*), (1967), Statistika, Moscow.

THE STATE AS A GEOGRAPHICAL FACTOR

Useful introductions to the working of the Soviet economy are:

CAMPBELL, R. W. (1967) *Soviet Economic Power: its organization, growth and challenge*, 2nd edn., Macmillan, London.
MILLER, M. (1965) *Rise of the Russian Consumer*, Institute of Economic Affairs, London.
NOVE, A. (1961) *The Soviet Economy*, Allen & Unwin, London.

CHAPTER 26

GEOGRAPHICAL STUDIES OF DEVELOPING AREAS: THE CASE OF TROPICAL AFRICA

A. M. O'CONNOR

CONCERN WITH TROPICAL AFRICA

During the past 20 years increasing attention has been paid in the economically advanced parts of the world to the problems of the less developed areas, and this interest has been reflected in increased geographical study of these areas, now politely termed "developing". In part this growth has only involved the application of well-established methods of teaching and research to new areas, rather than the setting of any new trend in aims or methods. In the 1950's the teaching of tropical African geography was greatly hindered by the lack of good textbooks on the continent, but during the 1960's a spate of new books has appeared, such as that by Mountjoy and Embleton (1965) which is most valuable as a reference work, that by Grove (1967) which is less detailed but more stimulating, and the outstanding text on African economic geography by Hance (1964). These books have met an urgent need, but most are quite conventional in approach.

Similarly, an increasing number of research papers on tropical Africa are appearing in European and American geographical journals, and also in several periodicals which have been established within tropical Africa, such as the *Nigerian Geographical Journal* and the *East African Geographical Review*. In most cases these papers also present the results of work using traditional techniques on new material. There is a continuing need for books and papers of this type, since many regions and topics have not as yet been closely studied and since many aspects of tropical Africa are rapidly changing. Yet at the same time there are certain new lines of approach which can be discerned.

NEW APPROACHES TO THE STUDY OF TROPICAL AFRICA

Some of the new trends which are more fully discussed elsewhere in this book are clearly relevant to the study of the developing countries. Any area can be studied either sub-regionally or systematically, but it is now widely considered that there is greater scope for advance in systematic studies than in traditional regional geography; and there are signs of a general swing in that direction, for example in university courses. As yet this is not very apparent in the geographical literature on tropical Africa, for most of the new teaching and reference texts examine the area country by country, and in some cases region by region within countries. However, the way was clearly shown some years ago by Kimble (1960), and the systematic approach has often been followed at the research level, as shown, for instance, in studies of migration by Prothero (1964), and of transport by Hance (1967).

An important new trend in teaching is towards interdisciplinary studies of developing areas, and the systematic approach links up well with study from the viewpoint of other disciplines. For the future economist or sociologist concerned with Africa the relevance of study of the economic and social geography of the continent is more obvious than that of work which covers its geography region by region. The differences in societies and in economic activities between one part of Africa and another are made more explicit through study by topics rather than regions, and this may more effectively counteract the systematic scientists' tendency to make sweeping generalizations about the whole of tropical Africa.

History plays a prominent part in most interdisciplinary area studies, but so far the past does not seem to command the same attention in work on tropical Africa as in parallel research on Asia and Latin America. There are now signs, however, of increasing recognition of the importance of an historical perspective for a full understanding of present geographical patterns. This is clearly demonstrated in the recent text by De Blij (1964), in the collection of essays on Africa in transition edited by Hodder and Harris (1967), and in Kay's work on patterns of settlement in Zambia (Kay, 1965). The need to consider past as well as present factors has increased now that differing colonial policies have joined pre-colonial conditions as matters of history. The transport network and the morphology of towns are but two of many features of tropical Africa which cannot be understood only in terms of the circumstances of today.

The general trend towards more quantitative study is clearly relevant for geography of developing areas, although much more often than in Europe and North America the data available are inadequate to justify the use of very refined statistical techniques. For example, there are several countries, such as Ethiopia and Somalia, which have never had a population census, and others, such as Nigeria, where census results are greatly disputed. There are very few countries with two reliable censuses from which changing patterns of population distribution can be studied quantitatively. Statistical techniques and theoretical models have been employed in a few books (e.g. Mabogunje,

1968), and in several research papers, such as that by Taaffe and others (1963) on transport networks in which certain regularities of pattern from one country to another were observed (see Chapter 16, above), and others in the physical branches of geography (e.g. Gregory, 1965). However, there is as yet nothing for tropical Africa comparable to the attempt made by Cole (1965) to use simple statistical methods for the analysis of various features in Latin America at a level within the reach of some sixth-form students.

A possible new approach in geography teaching is an effort to project the viewpoint of the people of the area concerned rather than that of outside observers. This sounds admirable in theory, but it is very difficult to recognize its implications in practice. There is as yet nothing written by African geographers that clearly demonstrates these implications for the study of tropical Africa, the nearest approach being the book by Ojo (1966) on the spatial aspects of his own people's culture. Most books and papers written by African geographers differ in no way from those written by outsiders, perhaps because they have all received a European type of training, but perhaps because there has not in fact been such an obvious European viewpoint in writings on African geography as there has been in the case of African history for example. Furthermore, if there is a distinct African view of African geography, that of a radical from Guinea may differ from that of a conservative from the Ivory Coast, and within any country that of the élite may not be the same as that of the peasantry. (Although South Africa lies outside the scope of this paper, it would be interesting to see how a geographical account by an African from within that country differed from those already written from the European side of the racial barrier.)

Probably the main implication of such a change of approach is a greater emphasis on differences within tropical Africa rather than differences between this region and Europe. The latter are now generally appreciated, but the former are still not always sufficiently recognized. Differences in the physical environment, in densities of population, and in patterns of economic activity are well within the traditional range of geographical study, yet even in relation to such matters as these this new focus is apparent. Thus attitudes towards the tropical environment have in the past swung from uninformed optimism to excessive pessimism, and only recently to a more widespread acceptance that it poses particularly severe problems in some areas, and presents particularly great opportunities in others. Similarly the significance of the contrasts between densely and sparsely populated countries for the choice between alternative lines of economic development is only gradually being fully appreciated.

Other differences between various parts of tropical Africa which are of vital importance for a full understanding of its geography, notably differences of social structures and political systems, have been given very little attention by geographers in the past. However, they occupy a prominent place in a few of the recent books mentioned above (e.g. Hodder and Harris, 1967), and the significance of these features is demonstrated in such research papers as those by Udo (1965) and Hunter (1967) on the dispersed settlement patterns found in parts of West Africa.

PROBLEM ORIENTATION

A current trend that is apparent in much geographical work, both systematic and regional, and that is relevant for school as well as university teaching, is towards the application of geography to practical problems (see Chapters 19–22 above). This is very evident in the study of developing areas, such as tropical Africa. Indeed, until recently they were normally described as "underdeveloped", a term which surely implied a problem to be tackled. This concern with the practical application of geography in these areas is demonstrated in a recent memorial volume to Sir Dudley Stamp in which seven essays are devoted to the developing world (Institute of British Geographers, 1968, pp. 145–258). Also indicative of this trend was the choice of the relevance of geography to problems of West African development as the theme for a recent inaugural lecture (Harrison Church, 1966). This certainly accords with the adoption of a local rather than a European viewpoint, for pure academic study is a luxury that Africa can scarcely afford.

In so far as attention is still directed to differences between Africa and Europe, there is a growing recognition that the most important difference is that of standards of living. The poverty of tropical Africa is still hardly mentioned in some geography texts, but in general much more consideration is now being given to this. The apparent tendency for the gap in incomes between richer and poorer parts of the world to widen is also sometimes discussed, although there is scope for much more work on this subject. The most useful recent contribution to the geography of underdevelopment, suitable for use in sixth forms as well as universities, is a book concerned with all the tropical lands by Hodder (1968). An earlier contribution made by Hance, which consisted of a number of African case studies, has recently been revised (Hance, 1967), and provides an excellent complement to Hodder's book.

The level of economic development is another field in which there are important differences between, and also within, tropical African countries. There are contrasts in the extent of the need for improvement between, for example, Ghana and Upper Volta, and even between southern and northern Ghana; and it is possible that on this scale also the contrast is at present increasing. In such cases a great problem is often posed by the decision on priorities between the areas in greatest need and those where the return on investment is likely to be highest. The ideal solution is perhaps the recognition of potential "growth poles" within the poorer parts of the region, where there is some prospect of development provided that a number of problems are tackled simultaneously, and this task presents the geographer with a great challenge.

Different countries, and the various districts within them, naturally differ in the types of development for which they are suited. This depends on, among other things, their physical resources, their demographic situation and their existing economic structures. Increasing attention is now being paid to the assessment of physical resources as an aspect of applied geography (Young, 1968), one example of this in tropical Africa being the evaluation of soils in Malawi by Brown and Young (1965). Attention has also been given to the applied aspects of population geography. Although population pressure such as is widespread in Asia occurs only locally in tropical Africa, work has been done in several of these localities, notably by Hunter (1967) in Ghana, by Mortimore (1967) in northern Nigeria, and by Ominde (1968), whose book on Kenya incorporates an intensive study of population pressures and consequent movements. As noted above, migration has also been studied by Prothero (1964 and 1965), whose work is a particularly clear example of applied geography since it was undertaken under the auspices of the World Health Organization and was concerned especially with migration as a factor hindering malaria control. Disease is, of course, one of the greatest problems of tropical Africa, and several geographers are giving attention to its spatial patterns.

Recent work on the economic geography of tropical Africa has included a number of studies focused on practical problems, such as the impact of new transport facilities (O'Connor, 1965) and new sources of power (Wilson, 1968). There is scope for further work of this kind in several countries following the opening of new railways, roads, and ports, and the establishment of new power stations; indeed such research is taking place in several university departments of geography in tropical Africa. Many geographers in the highly developed countries have applied their work to problems of town planning, and here again there are opportunities for useful work in tropical Africa, where urban development, while still generally slight, is now rapidly taking place. Examples of this have recently been provided on a broad scale by Harrison Church (1967), and for a particular West African city by Harvey and Dewdney (1968).

Geographers are also concerning themselves with the political problems of tropical Africa, and a recent collection (Fisher, 1968) included studies of political patterns in East Africa, of a persistent boundary problem in West Africa, and of the significance of the political map for transport in West Africa. Some of the political problems which have commanded much more attention in the world at large are also to some extent geographical in nature, since they involve conflicts between areas as well as between social groups. Examples include the struggle between Katanga and the central authorities in the Congo, that between Biafra and the central authorities in Nigeria, the opposition in southern Sudan to domination by the Arab north, and the claims of Somalia to the parts of neighbouring states that are occupied by Somali peoples. Another potential field for studies in applied geography is provided by the attempts that are being made to tackle the fundamental problem of underdevelopment by establishing closer co-operation between tropical African States. Little has yet been written by geographers on these topics, but attention is directed to them by Hodder and Harris (1967), and they are now frequently discussed in university courses on African geography. There is much that the sixth-form teacher can do with the assistance of the reports which appear in the daily and weekly press, which in Britain at least gives a prominent place to the political affairs of tropical Africa.

CONCLUSION

While various new trends can be discerned in the geographical study of developing areas such as tropical Africa, well-established methods of approach should not necessarily be discarded. It is still important that everyone in countries such as Britain should have some familiarity with the geography of the rest of the world, and that there should be some people with a specialist knowledge of particular areas; and for many years geography teachers have been achieving this. Yet, especially in relation to the developing countries, geography teaching can also highlight problems, and may contribute to the discussion of their solution. In this way students may become not only informed but also concerned about these lands. Only when the majority of the population of the developed countries feel such concern will there be any real prospect of lessening what is perhaps the most serious form of areal differentiation in the world, that between the rich lands and the poor.

REFERENCES AND FURTHER READING

The clearest indication of the nature of current geographical work on tropical Africa is provided by the recent literature. The references given here have been confined to reasonably accessible works by geographers, and mainly to those published since 1964. A thorough survey of earlier work has been provided by Steel in Steel, R. W. and Prothero. R. M. (Eds.) *Geographers and the Tropics*, Longmans, London, pp. 1–29. This list is also confined to books and papers in English, but it should be noted that the volume of literature in French continues to expand, notably in such journals as *Cahiers d'Outre Mer*, and that increased interest in African geography is also evident in Germany, the Soviet Union, and elsewhere.

CONCERN WITH TROPICAL AFRICA

The most recent standard texts on the geography of Africa, suitable for specialized sixth-form study or as an introduction to university courses, are:

GROVE, A. T. (1967) *Africa South of the Sahara*, Oxford U.P., London.
HARRISON CHURCH, R. J. *et al.* (1964) *Africa and the Islands*, Longmans, London.
MOUNTJOY, A. B. and EMBLETON, C. (1965) *Africa: a geographical study*, Hutchinson, London.
POLLOCK, N. C. (1968) *Africa*, U. of London P., London.

Another text, which selects certain themes for each country, is:

DE BLIJ, H. J. (1964) *A Geography of Subsaharan Africa*, Rand McNally, Chicago.

A very comprehensive study of African economic geography is:

HANCE, W. A. (1964) *The Geography of Modern Africa*, Columbia U.P., New York.

New geographical journals published in Africa (contributions to which have not been included individually in this list) include:

The Bulletin, the journal of the Sierra Leone Geographical Association.
Bulletin of the Ghana Geographical Association.
The East African Geographical Review.
The Ethiopian Geographical Journal.
The Nigerian Geographical Journal.

NEW APPROACHES TO THE STUDY OF TROPICAL AFRICA

The one notable geographical study of tropical Africa which adopts a systematic approach is:

KIMBLE, G. H. T. (1960) *Tropical Africa*, Twentieth Century Fund, New York.

Other books and research papers following a systematic approach include:

BARBOUR, K. M. and PROTHERO, R. M. (Eds.) (1961) *Essays on African Population*, Routledge & Kegan Paul, London.
HANCE, W. A. (1967) *African Economic Development*, Praeger, New York, chap. 5, Transport in tropical Africa, pp. 115–61.
HOYLE, B. S. and HILLING, D. (Eds.) (1969) *Seaports of Tropical Africa*, Macmillan, London.
PROTHERO, R. M. (1964) Continuity and change in African population mobility, in STEEL, R. W. and PROTHERO, R. M. (Eds.) *Geographers and the Tropics*, Longmans, London, pp. 189–214.
PROTHERO, R. M. (1965) *Migrants and Malaria*, Longmans, London.
THOMAS, M. F. and WHITTINGTON, G. W. (Eds.) (1969) *Environment and Land Use in Africa*, Methuen, London.

Books on more limited areas arranged systematically or confined to particular topics include:

CLARKE, J. I. (Ed.) (1966) *Sierra Leone in Maps*, U. of London P., London.
KAY, G. (1967) *A Social Geography of Zambia*, U. of London P., London.
LEBON, J. H. G. (1965) *Land Use in Sudan*, Geographical Publications, Bude.
MABOGUNJE, A. L. (1968) *Urbanization in Nigeria*, U. of London P., London.
MORGAN, W. B. and PUGH, J. C. (1969) *West Africa*, Methuen, London.
O'CONNOR, A. M. (1966) *An Economic Geography of East Africa*, Bell, London.
OMINDE, S. H. (1968) *Land and Population Movements in Kenya*, Heinemann, London.
PIKE, J. G. and RIMMINGTON, G. T. (1965) *Malawi*, Oxford U.P., London.

A book which emphasizes changing rather than static patterns is:

HODDER, B. W. and HARRIS, D. R. (Eds.) (1967) *Africa in Transition*, Methuen, London.

The results of research on changing patterns are presented in:

DICKSON, K. B. (1969) *A Historical Geography of Ghana*, Cambridge U.P., London.

KAY, G. (1965) *Changing Patterns of Settlement and Land Use in the Eastern Province of Northern Rhodesia*, U. of Hull P., Hull.

MCDONELL, G. (1964) The dynamics of geographic change: the case of Kano, *Annals of the Association of American Geographers*, **54**, 355–71.

UDO, R. K. and OGUNDANA, B. (1966), Factors influencing the fortunes of ports in the Niger delta, *Scottish Geographical Magazine* **82**, 169–83.

Examples of the application of statistical techniques and theoretical models to African situations are provided in:

GOULD, P. R. (1967) On the geographical interpretation of eigenvalues, *Transactions of the Institute of British Geographers*, **42**, 53–86.

GREGORY, S. (1965) *Rainfall over Sierra Leone*, U. of Liverpool P., Liverpool.

SOJA, E. W. (1968) *The Geography of Modernization in Kenya*, Syracuse U.P., Syracuse.

TAAFFE, E. J., MORRILL, R. L. and GOULD, P. R. (1963) Transport expansion in underdeveloped countries, *Geographical Review*, **53**, 503–29.

A textbook which demonstrates the use of such techniques for another underdeveloped area is:

COLE, J. P. (1965) *Latin America*, Butterworth, London.

One of the few advanced geographical studies which examines a part of tropical Africa from within is:

OJO, G. J. A. (1966) *Yoruba Culture, a Geographical Analysis*, U. of London P., London.

Geographical research emphasizing social factors is exemplified by:

FLOYD, B. and ADINDE, M. (1967) Farm settlements in Eastern Nigeria, *Economic Geography*, **43**, 189–230.

HUNTER, J. M. (1967) The social roots of dispersed settlement in northern Ghana, *Annals of the Association of American Geographers*, **57**, 339–49.

UDO, R. K. (1965) Disintegration of nucleated settlement in eastern Nigeria, *Geographical Review*, **55**, 55–67.

PROBLEM ORIENTATION

Attention is focused on the geography of underdevelopment in:

HANCE, W. A. (1967) *African Economic Development*, Praeger, New York.

HARRISON CHURCH, R. J. (1966) *Some Geographical Aspects of West African Development*, Bell, London.

HODDER, B. W. (1968) *Economic Development in the Tropics*, Methuen, London.

Institute of British Geographers (1968) *Land Use and Resources: studies in applied geography*, I.B.G. Special Publication, **1**.

MOUNTJOY, A. B. (1963) *Industrialization and Under-developed Countries*, Hutchinson, London.

A stimulating reconnaissance study of the spatial pattern of economic development though focused most sharply on southern Africa is:

GREEN, L. P. and FAIR, T. J. D. (1962) *Development in Africa*, Witwatersrand, U.P., Johannesburg.

The scope for applied physical geography in underdeveloped areas is indicated in:

MOSS, R. P. (Ed.) (1968) *The Soil Resources of Tropical Africa*, Cambridge U.P., Cambridge.

YOUNG, A. (1968) Natural resource surveys for land development in the tropics, *Geography* **53**, 229–48.

Specific examples are:

BROWN, P. and YOUNG, A. (1965) *The Physical Environment of Central Malawi*, Govt. Printer, Zomba.

KENWORTHY, J. M. (1964) Rainfall and the water resources of East Africa, in STEEL, R. W. and PROTHERO, R. M. (Eds.) *Geographers and the Tropics*, Longmans, London, pp. 111–37.

TURNER, B. J. and BAKER, P. R. (1968) Tsetse control and livestock development, a case study from Uganda, *Geography*, **53,** 249–59.

Examples of work in the field of applied population geography are:

HUNTER, J. M. (1967) Population pressure in a part of the West African savanna, *Annals of the Association of American Geographers*, **57,** 101–14.

MORTIMORE, M. J. (1967) Land and population pressure in the Kano close-settled zone, *Advancement of Science*, **118,** 699–86.

A number of geographers contributed to:

CALDWELL, J. C. and OKONJO, C. (1968) *The Population of Tropical Africa*, Longmans, London.

Studies in economic geography concerned with practical problems include:

BROOKE, C. (1967) Types of food shortage in Tanzania, *Geographical Review*, **57,** 333–57.

COPPOCK, J. T. (1966) Agricultural developments in Nigeria, *Journal of Tropical Geography*, **23,** 1–18.

GRIFFITHS, T. L. (1968) Zambian coal: an example of strategic resource development, *Geographical Review*, **58,** 538–51.

HOYLE, B. S. (1967) *The Seaports of East Africa*, East African Publishing House, Nairobi.

MAY, J. M. (1965) *The Ecology of Malnutrition in Middle Africa*, Hafner, New York.

O'CONNOR, A. M. (1965) *Railways and Development in Uganda*, Oxford U. P., Nairobi.

PETEREC, R. J. (1967) *Dakar and West African Economic Development*, Columbia U. P., New York.

UDO, R. K. (1965) Problems of developing the Cross River District of Eastern Nigeria, *Journal of Tropical Geography*, **20,** 65–74.

WILSON, G. G. (1968) *Owen Falls*, East African Publishing House, Nairobi.

Several geographers contributed to:

RUBIN, N. and WARREN, W. M. (1968) *Dams in Africa*, Cass, London.

Geographical contributions to African town planning problems are:

HARRISON CHURCH, R. J. (1967) Urban problems and economic development in West Africa, *Journal of Modern African Studies*, **5,** 511–20.

HARVEY, M. and DEWDNEY, J. (1968) Planning problems in Freetown, in FYFE, C. and JONES, E. (Eds.) *Freetown: a symposium*, Sierra Leone U.P., Freetown, pp. 179–95.

Three papers on problems of African political geography are to be found in:

FISHER, C. A. (Ed.) (1968) *Essays in Political Geography*, Methuen, London, which is largely concerned with the new states of the developing world.

INDEX

285